Cambridge Elements ≡

Elements in Applied Category Theory
edited by
Bob Coecke
Cambridge Quantum Ltd
Joshua Tan
University of Oxford

THEORETICAL COMPUTER SCIENCE FOR THE WORKING CATEGORY THEORIST

Noson S. Yanofsky
Brooklyn College, City University of New York

CAMBRIDGE
UNIVERSITY PRESS

CAMBRIDGE
UNIVERSITY PRESS

University Printing House, Cambridge CB2 8BS, United Kingdom

One Liberty Plaza, 20th Floor, New York, NY 10006, USA

477 Williamstown Road, Port Melbourne, VIC 3207, Australia

314–321, 3rd Floor, Plot 3, Splendor Forum, Jasola District Centre,
New Delhi – 110025, India

103 Penang Road, #05–06/07, Visioncrest Commercial, Singapore 238467

Cambridge University Press is part of the University of Cambridge.

It furthers the University's mission by disseminating knowledge in the pursuit of
education, learning, and research at the highest international levels of excellence.

www.cambridge.org
Information on this title: www.cambridge.org/9781108792745
DOI: 10.1017/9781108872348

First published 2022

A catalogue record for this publication is available from the British Library.

ISBN 978-1-108-79274-5 Paperback
ISSN 2633-1861 (online)
ISSN 2633-1853 (print)

Theoretical Computer Science for the Working Category Theorist

Elements in Applied Category Theory

DOI: 10.1017/9781108872348
First published online: January 2022

Noson S. Yanofsky
Brooklyn College, City University of New York
Author for correspondence: Noson S. Yanofsky,
noson@sci.brooklyn.cuny.edu

Abstract: Using basic category theory (category, functor, natural transformation, etc.), this Element describes all the central concepts and proves the main theorems of theoretical computer science. Category theory, which works with functions, processes, and structures, is uniquely qualified to present the fundamental results of theoretical computer science. In this text, we meet some of the deepest ideas and theorems of modern computers and mathematics, e.g., Turing machines, unsolvable problems, the P=NP question, Kurt Gödel's incompleteness theorem, intractable problems, Turing's Halting problem, and much more. The concepts come alive with many examples and exercises. This short text covers the usual material taught in a year-long course.

Keywords: category theory, computational models, computability theory, complexity theory

MSC2020 Subject Classification: 18AXX, 18DXX, 18MXX, 03BXX, 03DXX, 68NXX, 68QXX, 68RXX

ISBNs: 9781108792745 (PB), 9781108872348 (OC)
ISSNs: 2633-1861 (online), 2633-1853 (print)

Dedicated to the memory
of my brother
Eli

This work is dedicated to the memory of my brother Rabbi Eliyahu Mordechai Yanofsky ZTL (February 19, 1964 – March 12, 2018). He was a brilliant scholar who spent his entire life studying Jewish texts and helping others. Eli had a warm personality and a clever wit that endeared him to many. He was an inspiration to hundreds of friends and students. As an older brother, he was a paragon of excellence. He left behind parents, a wife, siblings, and ten wonderful children. His loss was terribly painful. He is sorely missed.

Contents

Preface   1

1 Introduction   4

2 Aide-Mémoire on Category Theory   9

3 Models of Computation   16

4 Computability Theory   60

5 Complexity Theory   77

6 Diagonal Arguments   104

7 Conclusion   125

References   130

Preface

"Unsolvable problems," "The P=NP question," "Alan Turing's Halting problem," "Kurt Gödel's incompleteness theorem," "Intractable problems," etc. These are just some of the many buzzwords that one hears about in the current technology media. These concepts are all associated with theoretical computer science which touches on some of the deepest and most profound parts of contemporary science and mathematics. The issues discussed in this work emerge in almost every area of computers, theoretical physics, and mathematics. In this short text you will learn the main ideas and theorems of theoretical computer science.

While there are many textbooks about theoretical computer science, this Element is unique. It is the first text that translates the ideas of theoretical computer science into the basic language of category theory. With some of the simple concepts of category theory in hand, the reader will be able to understand all the ideas and theorems that are taught in a year-long course of theoretical computer science.

Why is it better to learn theoretical computer science with category theory?

- It is easy! Most of the ideas and theorems of theoretical computer science are consequences of the categorical definitions. Once the categories and functors of the different models of computation are set up, the concepts of theoretical computer science simply emerge. Many theorems are a straightforward consequence of composition and functoriality.
- It is short! The powerful language of category theory ensures us that what is usually taught in a large textbook can be taught in these few pages.
- It gives you the right perspective! Anyone who has ever experienced any category theory knows that it makes you "zoom out" and see things from a "bird's-eye view." Rather than getting overwhelmed by all the little details so that you cannot "see the forest for the trees," category theory lets you see the big picture.
- It gets to the core of the issues! Category theory distills the important aspects of a subject and leaves one with the main structures needed to understand what is going on.
- It connects better with other areas of science and math! Category theory has become the *lingua franca* of numerous areas of mathematics, and theoretical physics (e.g., Baez and Stay, 2011; Coecke and Kissinger, 2017). By putting the ideas of theoretical computer science in the language of category theory, connections are made to these other areas.

There are several texts and papers for the computer scientist to learn category theory (Barr and Wells, 1999; Walters, 1991; Pierce, 1991; Asperti and Longo, 1991; Scott, 2000; Spivak, 2014; Poigné, 1992). Ours is totally different. The intended audience here is someone who already knows the basics of category theory and who wants to learn theoretical computer science.

One does not need to be a category theory expert to read this Element. We do not assume that the reader knows much more than the notion of category, functor, and equivalence. Any more advanced categorical concepts are taught in §2 or "on the fly" when needed.

Organisation

We start with a quick introduction to the major themes of theoretical computer science and explain why category theory is uniquely qualified to describe those ideas and theorems. The next section is a short teaching aid of some categorical structures that arise in the text and that might be unknown to the category theory novice.

Since we will be discussing computation, we must fix our ideas and deal with various *models of computation*. What do we mean by a computation? The literature is full of such models. Section 3 classifies and categorizes these various models while showing how they are all linked together. With the definitions and notation given in §§3.1 and 3.2, one can safely move on to almost any other part.

Sections 4 and 5 are the core of the work. Section 4 describes which functions can be computed by a computer – and more importantly – which functions *cannot* be computed by a computer. This is called *computability theory*. We give many examples of functions that no computer can perform. Along the way, we meet Alan Turing's Halting problem and Kurt Gödel's famous incompleteness theorem. We also discuss a classification and hierarchy of all the functions that cannot be computed by computers.

We go on to discuss *complexity theory* in §5. This is where we ask and answer what is *efficiently* computable. Every computable function demands a certain amount of time or space to compute. We discuss how long and how complicated certain computations are. Along the way we meet the famous NP-complete problems and the P=NP question.

Section 6 is about a special type of proof that arises in many parts of theoretical computer science (and many areas of mathematics) called a *diagonal argument*. Although the diagonal arguments come in many different forms, we show that they are all instances of a single simple theorem of category theory. Along the way we meet Stephen Kleene's recursion theorem, Georg Cantor's different levels of infinity, and John von Neumann's self-reproducing machine.

Section 6 summarizes what was learned about theoretical computer science from the categorical perspective. It lists off some common themes that were seen throughout the text. It also has a guide for readers who wish to continue with their studies.

While computability theory and complexity theory are the main topics taught in a typical theoretical computer science class, there are many other topics that are either touched on or are taught in more advanced courses. A supplementary section on the web page of the text is a series of short introductions for several exciting topics: *Formal language theory*, *Cryptography*, *Kolmogorov complexity theory*, and *Algorithms*.

The text has many examples and exercises. The reader is strongly urged to work out all the exercises. In addition, the Element is sprinkled with short "Advanced Topics" that point out certain advanced theorems or ideas. We also direct the reader to where they can learn more about these topics. Many sections end with ideas for a potential "Research Project."

At the end of every section, the reader is directed to places where they can learn more about the topic from sources in classical theoretical computer science and in category theory.

There are, however, omissions for reasons of space. We do not cover every part of theoretical computer science. For example, we will not deal with program semantics and verification, analysis of algorithms, data structures, and information theory. While all these are interesting and can be treated with a categorical perspective, we have omitted these topics because of space considerations. Invariably, even within the topics that we do cover, certain theorems and ideas are missed or glossed over.

This is not a textbook about the relationship between category theory and computer science. Over the past half century, category theorists have made tremendous advances in computer science by applying categorical concepts and constructions to the structures and processes of computation. The literature in this area is immense. Section 7.2 gives some texts and papers to look for more in this direction. This text will not present all these areas. Rather, it is focused on the task at hand. Here we use a novel presentation to understand the classic parts of theoretical computer science.

This text does not stand alone. I maintain a web page for the text at
  http://www.sci.brooklyn.cuny.edu/ noson/TCStext.html
The web page contains supplementary material. There will also be links to interesting books and articles on category theory and theoretical computer science, some solutions to exercises, and a list of errata. The reader is encouraged to send any and all corrections and suggestions to
  noson@sci.brooklyn.cuny.edu.

1 Introduction

Theoretical computer science started before large-scale electronic computers actually existed. In the 1930s, when engineers were just beginning to work out the problems of making viable computers, Alan Turing and others were already exploring the limits of computation. Before physicists and engineers began struggling to create quantum computers, theoretical computer scientists designed algorithms for quantum computers and described their limitations. Even today, before there are any large-scale quantum computers, theoretical computer science is working on "post-quantum cryptography." The prescient nature of this field is a consequence of the fact that it studies only the important and foundational issues about computation. It asks and answers questions such as "What is a computation?" "What is computable?" "What is efficiently computable?" "What is information?" "What is random?" "What is an algorithm?" etc.

For us, the central role of a computer is to calculate functions. Computers input data of a certain type, manipulate the data, and have outputs of a certain type. In order to study computation we must look at the collection of all such functions. We must also look at all methods of computation and see how they describe functions. To every computational method, there is an associated function. What will be important is to study which functions are computed by a computational method and which are not. Which functions are easily computed and which functions need more resources? All these issues – and many more – will be dealt with in these pages.

Category theory is uniquely qualified to present the ideas of theoretical computer science. The basic language of category theory was formulated by Samuel Eilenberg and Saunders Mac Lane in the 1940s. They wanted to classify and categorize topological objects by associating them to algebraic objects. To do this they had to formulate a language that was not specifically related to topology or algebra. It needed to be abstract enough to deal with both topological and algebraic objects. This is where category theory gets its power. By being about nothing in particular, or "general abstract nonsense," it is about everything. In this text, we will see categories containing various types of functions and different models of computation. There are then functors comparing these functions and computational methods.

When categories first started, the morphisms in a category were thought of as homomorphisms between algebraic structures or continuous maps between topological spaces. The morphism $f: X \longrightarrow Y$ might be a homomorphism f from group X to group Y or it might be a continuous map from topological space X to topological space Y. However, as time progressed, researchers realized that they might look at f as a *process* of going from X to Y. An algebraic example is when X and Y are vector spaces and f is a linear transformation, a way of going from X to Y. In logic, X and Y can be propositions and f is a way of showing that after assuming X one can infer Y. Or for logicians interested in proofs, again X and Y were propositions and f represented a proof (or a formal way of showing) that with X as an assumption, one can conclude with Y. Physicists also got into the action. For them, X and Y were states of a system and f represented a dynamic or a process of going from one to the other. Geometers and physicists considered X and Y to be manifolds of a certain type and f to be a larger shape whose boundaries were X and Y. Computer scientists consider X and Y to be types of data and f is a function or a computational process that takes inputs of type X to output of type Y. In this text, many of our categories will be of this form. (See Baez and Stay, 2011, Coecke and Kissinger, 2017, for comparisons of many different types of processes.)

It pays to reexamine the definition of a category with the notion of a morphism as a process. When there are two composable maps,

$$X \xrightarrow{\ \ f\ \ } Y \xrightarrow{\ \ g\ \ } Z, \qquad (1.1)$$

their composition $g \circ f$ is to be thought of as *sequential processing*, i.e., first performing the f process and then performing the g process. If X is any object, then there is an identity process $\mathrm{Id}_X : X \longrightarrow X$ which does nothing to X. No change is made. When any process is composed with the identity process, it is essentially the same as the original process. When there are three composable maps,

$$X \xrightarrow{\ \ f\ \ } Y \xrightarrow{\ \ g\ \ } Z \xrightarrow{\ \ h\ \ } W, \qquad (1.2)$$

the associativity $(h \circ g) \circ f = h \circ (g \circ f)$ essentially follows from the fact that we are doing these processes in order: first f, then g, and finally h.

In the 1960s, with the development of *monoidal categories*, it was realized that not only do categories contain structures, but sometimes categories themselves have structure. In certain cases objects and morphisms of categories can be "tensored" or "multiplied." For example, given X and X' in the category of vector spaces, one can form their tensor product $X \otimes X'$ as an object of the category of vector spaces. Or, if X and X' are objects in the category of sets, then their Cartesian product, $X \times X'$, is also an object in the category of sets. As usual in category theory, we are concerned with morphisms. If $f : X \longrightarrow Y$ and $f' : X' \longrightarrow Y'$ are morphisms then we cannot only form $X \otimes X'$ and $Y \otimes Y'$

but also $f \otimes f'$ which we can write as

$$
\begin{array}{ccc}
X \xrightarrow{\quad f \quad} Y & & \\
\otimes & \text{or} & X \otimes X' \xrightarrow{\quad f \otimes f' \quad} Y \otimes Y'. \\
X' \xrightarrow{\quad f' \quad} Y' & &
\end{array} \tag{1.3}
$$

This should be considered as modeling *parallel processing*, i.e., performing the f and f' process independently and at the same time. A category where one can tensor the objects and morphisms is called a monoidal category. When the $f \otimes f'$ process is essentially the same as the $f' \otimes f$ process, we call the structure a *symmetric monoidal category*. Most of the categories we will meet in these pages will be types of symmetric monoidal categories.

Let us summarize. The categories we will use have objects that are types of data that describe the inputs and outputs of functions and processes, while the morphisms are the functions or the processes. There will be two ways of composing the morphisms: (i) regular composition will correspond to function composition or sequential processing, and (ii) the monoidal composition will correspond to parallel processing of functions or processes. There will be various functors between such categories which take computational procedures to functions or other computational procedures. We will ask questions such as: When are these functors surjective? What is in their image? What is not in their image? When are they equivalences? etc.

This Element will not only use the language of categories, it will also use the mindset of categories. As anyone who has studied a topic in category theory knows, this language has a unique way of seeing things. Rather than jumping in to the nitty-gritty details, categorists insist on setting up what they are talking about first. They like to examine the larger picture before getting into the details. Also, category theory has a knack for getting to the essence of a problem while leaving out all the dross that is related to a subject. Theoretical computer science stands to gain from this mindset.

Here are some of the topics that we will meet in these pages. We start off by considering different *models of computation*. These are formal, virtual machines that perform computations. We classify them into three different groupings. There are models, such as *Turing machines*, that perform symbol manipulations. Historically, this is one of the first computational models. There are models called *register machines* that perform computations through manipulating whole numbers. A hand-held calculator is a type of computer that manipulates numbers. Yet another computational model is called *circuits* and performs bit manipulations. Every modern digital electronic computer works by manipulating bits. And finally, another way of describing computation

– which is not really a computational model – is with *logical formulas*. For each of these types of computational model, we will describe a categorical structure that contains the models. There are functors from the categories of models to the categories of functions. There also exist functors between the categorical structures of models. For example, a hand-held calculator appears to manipulate whole numbers. In fact, what is really happening is that there is a circuit which manipulates bits that performs the calculations. This is the essence behind a functor between the categorical structure of register machines and the categorical structure of circuits.

Let us look at theoretical computer science from a broadly philosophical perspective. By the "syntax" of a function we mean a description of the function such as a program that implements the function, a computer that runs the function, a logical formula that characterizes the function, a circuit that executes the function, etc. By the "semantics" of a function we mean the rule that assigns to every input an output. Computability theory and complexity theory study the relationship between the syntax and the semantics of functions. There are many aspects to this relationship. Computability theory asks what functions are defined by syntax, and – more interestingly – what functions are *not* defined by syntax. Complexity theory asks how can we classify and characterize functions by examining their syntax.

In a categorical setting, the relationship between the syntax and semantics of functions is described by a functor from a category of syntax to a category of semantics. The functor takes a description of a function to the function it describes. Computability theory then asks what is in the image of this functor and – more interestingly – what is *not* in the image of the functor. Complexity theory tries to classify and characterize what is in the image of the functor by examining its preimage.

Let us look at some of the other topics we cover on these pages and the supplement. Diagonal arguments are related to self-reference. About two and a half thousand years ago, Epimenides (a cantankerous Cretan philosopher who proclaimed Cretans always lie) taught us that language can talk about itself and is self-referential. He showed that because of this self-reference there is a certain limitation of language (some sentences are true if and only if they are false.) In the late nineteenth century, the German mathematician Georg Cantor showed that sets can be self-referential and Bertrand Russell showed that one must restrict this ability or set theory will be inconsistent. In the 1930s, Kurt Gödel showed that mathematical statements can refer to themselves ("I am unprovable") and hence there is a limitation to the power of proofs. We are chiefly interested in Alan Turing showing that computers can reference themselves (after all, an operating system is a program dealing with programs).

This entails a limitation of computers. Diagonal arguments are used in systems with self-reference to find some limitation of the system. We describe a simple categorical theorem that models all these different examples of self-reference. Many results of theoretical computer science are shown to be instances of this theorem.

Formal language theory deals with the interplay between machines and languages. The more complicated a machine, the more complicated is the language it understands. With categories, we actually describe a functor from a category of machines to a category of languages. Machines that are weaker than Turing machines and their associated languages are explored.

Modern cryptography is about using computers to encrypt messages. While it should be computationally easy to encode these messages, only the intended receiver of the message should be able to decode with ease. Categorically, we will use some of the ideas we learned in complexity theory to discuss a subcategory of computable functions containing easily computable functions. In contradistinction, some other computable functions can be thought of as hard. A good cryptographic protocol is when the encoding and decoding are easy to compute and where the decoder is hard to find. The power of category theory will become apparent when we make a simple categorical definition and show that most modern cryptographic protocols can be seen an instance of this definition.

Kolmogorov complexity theory is concerned with the complexity of strings. We use Turing machines to classify how complicated strings of characters are. The section also has a discussion of the notion of randomness.

Algorithms are at the core of computer science. From a broad prospective, algorithms are on the thin line between the syntax and semantics of computable functions. The functor from syntax to semantics factors as

$$\text{Syntax} \longrightarrow \text{Algorithms} \longrightarrow \text{Semantics}. \tag{1.4}$$

Or we can see it as

$$\text{Programs} \longrightarrow \text{Algorithms} \longrightarrow \text{Functions}. \tag{1.5}$$

This view affords us a deeper understanding of the relationship between programming, computer science, and mathematics.

The first part of §7 is a continuation of this Introduction. It summarizes what was done in this text using technical language. We show that there are certain unifying themes and categorical structures that make theoretical computer science more understandable. We also close that section with a reader's guide to going further with their studies.

Let's roll up our sleeves and get to work!

2 Aide-Mémoire on Category Theory

We assume the basics of category theory. The reader is expected to know the concept of a category, functor, natural transformation, equivalence, limit, etc. We review some less well-known concepts of category theory. Any other required categorical constructions will be reviewed along the way. Feel free to skip this section and return as needed.

2.1 Slice Categories and Comma Categories

The slice and coslice constructions make the morphisms of one category into the objects of another category.

Definition 2.1. Given a category A and an object a of A, the *slice category*, A/a, is a category whose objects are pairs $(b, f: b \longrightarrow a)$ where b is an object of A and f is a morphism of A whose codomain is a. A morphism of A/a from $(b, f: b \longrightarrow a)$ to $(b', f': b' \longrightarrow a)$ is a morphism $g: b \longrightarrow b'$ of A that makes the following triangle commute:

$$(2.1)$$

Composition of morphisms comes from the composition in A, and the identity morphism for the object $(b, f: b \longrightarrow a)$ is id_b.

Example 2.2. Some examples of a slice category:

- Consider the category **Set**. Let $\mathbb{R}$ be the set of all real numbers. The category **Set**/$\mathbb{R}$ is the collection of all $\mathbb{R}$-valued functions.
- Let us foreshadow by showing an example that we will see in our text. Let **Func** be the category whose objects are different types and whose morphisms are all functions between types. One of the given types is Boolean, Bool, which has the possible values of 0 and 1. Then **Func**/Bool is the category whose objects are functions from any type to Bool. Such functions are called *decision problems* and they will play a major role in the coming pages. We also have a notion of a morphism between decision problems:

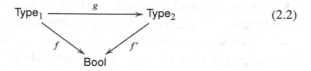

$$(2.2)$$

where f and f' are decision problems and g is a morphism from f to f'. Such a morphism of decision problems will be called a "reduction." The map g is a reduction of f to f'. The idea behind a reduction is that if we know the answer to f' then we can also find the answer to f. Since the triangle commutes, to find the answer to f on an input, simply evaluate the input using g and then find the answer using f'. In other words, solving f reduces to solving f'.

Exercise 2.3. Show that $\mathrm{id}_a : a \longrightarrow a$ is the terminal object in $\mathbf{A}/a$.

Exercise 2.4. Prove that any $f : a \longrightarrow a'$ in $\mathbf{A}$ induces a functor $f_* : \mathbf{A}/a \longrightarrow \mathbf{A}/a'$ that takes any object $g : b \longrightarrow a$ of $\mathbf{A}/a$ to $f \circ g$ is in $\mathbf{A}/a'$.

The dual notion of a slice category is a coslice category:

Definition 2.5. Given a category $\mathbf{A}$ and an object a of $\mathbf{A}$, the *coslice category*, $a/\mathbf{A}$ is a category whose objects are pairs $(b, f : a \longrightarrow b)$ where b is an object of $\mathbf{A}$ and f is a morphism of $\mathbf{A}$ with domain a. A morphism of $a/\mathbf{A}$ from $(b, f : a \longrightarrow b)$ to $(b', f' : a \longrightarrow b')$ is a morphism $g : b \longrightarrow b'$ of $\mathbf{A}$ that makes the following triangle commute:

$$(2.3)$$

Composition of morphisms comes from the composition in $\mathbf{A}$, and the identity morphism for the object $(b, f : b \longrightarrow a)$ is id_b.

Example 2.6. Some examples of a coslice category:

- Consider the one element set $\{*\}$. The category $\{*\}/\mathbf{Set}$ has objects that are sets with a function that picks out a distinguished element of the set. An object is a pair (S, s_0) where S is a set and s_0 is a distinguished element of that set. The morphisms from (S, s_0) to (T, t_0) are set functions that preserve the distinguished element. That is, $f : S \longrightarrow T$ such that $f(s_0) = t_0$. This is the category of pointed sets.
- When we talk about finite automata, we will generalize the above example. Let **FinGraph** be the category of finite graphs and graph homomorphisms. Let $\mathbf{2_0}$ be the graph with just two vertices (named s and t for "source" and "target.") Then $\mathbf{2_0}/\mathbf{FinGraph}$ is the category of finite graphs with a distinguished vertex for s and a distinguished vertex for t. (They might be the same vertex.)
- Let **Prop** be the category of propositions with at most one morphism from a to b if a entails b. Let $p \in \mathbf{Prop}$ be a particular proposition. Then $p/\mathbf{Prop}$ is the category of all propositions that are implied by p.

Exercise 2.7. Show that $\mathrm{id}_a : a \longrightarrow a$ is the initial object in $a/\mathbf{A}$.

Definition 2.8. Given functors $F: \mathbf{A} \longrightarrow \mathbf{C}$ and $G: \mathbf{B} \longrightarrow \mathbf{C}$, one can form the *comma category* (F, G), sometimes written $F \downarrow G$. The objects of this category are triples (a, f, b) where a is an object of $\mathbf{A}$, b is an object of $\mathbf{B}$ and, $f: F(a) \longrightarrow G(b)$ is a morphism in $\mathbf{C}$. A morphism from (a, f, b) to (a', f', b') in (F, G) consists of a pair of morphisms (g, h) where $g: a \longrightarrow a'$ is in $\mathbf{A}$ and $h: b \longrightarrow b'$ is in $\mathbf{B}$ such that the following diagram commutes:

$$
\begin{array}{ccc}
F(a) & \xrightarrow{\ \ f\ \ } & G(b) \\
{\scriptstyle F(g)}\downarrow & & \downarrow{\scriptstyle G(h)} \\
F(a') & \xrightarrow[\ \ f'\ \ ]{} & G(b').
\end{array}
\tag{2.4}
$$

Composition of morphisms comes from the fact that one commuting square on top of another ensures that the whole diagram commutes. The identity morphisms are obvious.

Example 2.9. Some examples of comma categories are familiar already.

- If $\mathbf{B} = \mathbf{1}$ and $G: \mathbf{1} \longrightarrow \mathbf{C}$ picks out the object c_0 and $\mathbf{A} = \mathbf{C}$ with $F = \mathrm{Id}_{\mathbf{C}}$, then the comma category (F, G) is the slice category $\mathbf{C}/c_0$.
- If $\mathbf{A} = \mathbf{1}$ and $F: \mathbf{1} \longrightarrow \mathbf{C}$ picks out the object c_0 and $\mathbf{B} = \mathbf{C}$ with $G = \mathrm{Id}_{\mathbf{C}}$, then the comma category (F, G) is the coslice category $c_0/\mathbf{C}$.

In this text, we will use a comma category construction with F being an inclusion functor. In this case the comma category will have objects of a certain type and morphisms from the subcategory. This will be important for reductions of one problem to another.

There are forgetful functors $U_1: (F, G) \longrightarrow \mathbf{A}$ and $U_2: (F, G) \longrightarrow \mathbf{B}$ which are defined on objects as follows: U_1 takes (a, f, b) to a and U_2 takes (a, f, b) to b.

Exercise 2.10. Show that if the following two triangles of categories and functors commute

$$
\begin{array}{ccc}
\mathbf{A} & \xrightarrow{\ \ F'\ \ } & \mathbf{C}' \\
{\scriptstyle F}\downarrow & \nearrow & \uparrow{\scriptstyle G'} \\
\mathbf{C} & \xleftarrow[\ \ G\ \ ]{} & \mathbf{B}
\end{array}
\tag{2.5}
$$

where $\mathbf{C} \hookrightarrow \mathbf{C}'$ is an inclusion functor, then there is an inclusion functor $(F, G) \hookrightarrow (F', G')$.

2.2 Symmetric Monoidal Categories and Bicategories

Many of our categories are collections of processes where, besides sequentially composing processes, we can compose processes in parallel. Such categories are formulated as categories with extra structure.

Definition 2.11. A *strictly associative monoidal category* or a *strict monoidal category* is a triple $(\mathbf{A}, I, \otimes)$ where $\mathbf{A}$ is a category, I is an object in $\mathbf{A}$ called a "unit," and $\otimes$ is a bifunctor $\otimes \colon \mathbf{A} \times \mathbf{A} \longrightarrow \mathbf{A}$ called a "tensor" which satisfies the following axioms:

- $\otimes$ is strictly associative, i.e., for all objects a, b and c in $\mathbf{A}$, $(a \otimes b) \otimes c = a \otimes (b \otimes c)$, and
- I acts like a unit, i.e., for all objects a in $\mathbf{A}$, $a \otimes I = a = I \otimes a$.

Example 2.12. Any monoid (including $\mathbb{N}$, the natural numbers) can be thought of as a strictly associative monoidal category. The elements of the monoid form a discrete category and the unit of the monoid becomes the unit of the symmetric monoidal category. The multiplication in the monoid becomes the tensor of the category.

In the literature, there is a weaker definition of a *monoidal category*. This is a category where there is an isomorphism between $(a \otimes b) \otimes c$ and $a \otimes (b \otimes c)$. These isomorphisms must satisfy higher-dimensional axioms called a "coherence conditions." While this concept is very important in many areas of mathematics, physics, and computer science, it will not arise in our presentation. There is a theorem of Saunders Mac Lane that says that every monoidal category is equivalent (in a very strong way) to a strict monoidal category. The categories that we will be dealing with will have the natural numbers or sequences of types as objects and will be strict monoidal categories.

Before we go on, we need the following functor. For any category $\mathbf{A}$, there is a "twisting" functor

$$\mathrm{tw} \colon A \times \mathbf{A} \longrightarrow \mathbf{A} \times \mathbf{A} \tag{2.6}$$

that is defined for two elements a and b as $\mathrm{tw}(a, b) = (b, a)$. The functor tw is defined similarly for morphisms.

Definition 2.13. A *symmetric strictly monoidal category* or *symmetric monoidal category* is a quadruple $(\mathbf{A}, I, \otimes, \gamma)$ where $(\mathbf{A}, I, \otimes)$ is a strict monoidal category and γ is a natural transformation that is an isomorphism $\gamma \colon \otimes \longrightarrow (\otimes \circ \mathrm{tw})$, i.e., for all objects a, b in $\mathbf{A}$, there is an isomorphism $a \otimes b \xrightarrow{\ \gamma_{a,b}\ } b \otimes a$ which is natural in a and b. γ must satisfy the following coherence axioms:

- the symmetry axiom,

$$\gamma_{b,a} \circ \gamma_{a,b} = \mathrm{Id}_{a\otimes b} \tag{2.7}$$

- the braiding axiom,

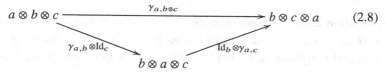 (2.8)

This γ is called a "braiding."

Of course we are not only interested in such symmetric monoidal categories, but in the way they relate to each other.

Definition 2.14. A *strong symmetric monoidal functor* from $(\mathbf{A}, I, \otimes, \gamma)$ to $(\mathbf{A}', I', \otimes', \gamma')$ is a triple (F, ι, ∇) where $F: \mathbf{A} \longrightarrow \mathbf{A}'$ is a functor, $\iota: I' \longrightarrow F(I)$ is an natural isomorphism in $\mathbf{A}'$, and $\nabla: (\otimes' \circ (F \times F)) \longrightarrow (F \circ \otimes)$ is a natural transformation which is an isomorphism. That is, for every a, a' in $\mathbf{A}$, there is an isomorphism $\nabla_{a,a'}: F(a) \otimes' F(a') \longrightarrow F(a \otimes a')$ natural in a and a'. These natural transformations must satisfy the following coherence rules

- ∇ must cohere with the braidings

$$
\begin{array}{ccc}
F(a) \otimes' F(a') & \xrightarrow{\nabla_{a,a'}} & F(a \otimes a') \\
{\scriptstyle \gamma'_{F(a),F(a')}}\downarrow & & \downarrow{\scriptstyle F(\gamma_{a,a'})} \\
F(a') \otimes' F(a) & \xrightarrow{\nabla_{a',a}} & F(a' \otimes a);
\end{array}
\tag{2.9}
$$

- ι must cohere with ∇

$$
\begin{array}{ccc}
I' \otimes' F(a) & \xrightarrow{\iota \otimes' Id} & F(I) \otimes' F(a) \\
\| & & \downarrow{\scriptstyle \nabla_{I,a}} \\
F(a) & =\!=\!=\!=\!= & F(I \otimes a);
\end{array}
\tag{2.10}
$$

(there is a similar coherence requirement with ι and $\nabla_{a,I}$;)
- ∇ must cohere with itself, i.e., it is associative

$$
\begin{array}{ccc}
F(a) \otimes' F(a') \otimes' F(a'') & \xrightarrow{Id \otimes' \nabla_{a',a''}} & F(a) \otimes' F(a' \otimes a'') \\
{\scriptstyle \nabla_{a,a'} \otimes' Id}\downarrow & & \downarrow{\scriptstyle \nabla_{a,a'\otimes a''}} \\
F(a \otimes a') \otimes' F(a'') & \xrightarrow{\nabla_{a\otimes a',a''}} & F(a \otimes a' \otimes a'').
\end{array}
\tag{2.11}
$$

We will also need a stronger notion of a functor.

Definition 2.15. A *symmetric monoidal strict functor* from $(\mathbf{A}, I, \otimes, \gamma)$ to $(\mathbf{A}', I', \otimes', \gamma')$ is a functor $F: \mathbf{A} \longrightarrow \mathbf{A}'$ where $F(I) = I'$ and $F(a) \otimes' F(a') = F(a \otimes a')$; i.e., the ι and ∇ are the identity.

Definition 2.16. A *symmetric monoidal natural transformation* from (F, ι, ∇) to (F', ι', ∇') is a natural transformation $\mu: F \longrightarrow F'$; i.e., for every a in $\mathbf{A}$, a natural morphism $\mu_a: F(a) \longrightarrow F'(a)$ which must satisfy the following coherence axioms:

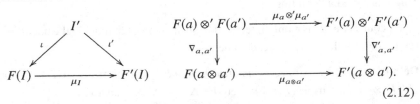

$$(2.12)$$

Some of these axioms will easily be satisfied when we deal with strict functors.

Given the notion of a symmetric monoidal natural transformation, we can easily describe what it means for two symmetric monoidal categories to be equivalent.

Definition 2.17. There is a *symmetric monoidal equivalence* between $(\mathbf{A}, I, \otimes, \gamma)$ and $(\mathbf{A}', I', \otimes', \gamma')$ if there is a symmetric monoidal functor (F, ι, ∇) from $\mathbf{A}$ to $\mathbf{A}'$ and a symmetric monoidal functor (F', ι', ∇') from $\mathbf{A}'$ to $\mathbf{A}$ with symmetric monoidal natural transformations which are isomorphisms $\mu: \mathrm{Id}_\mathbf{A} \longrightarrow F' \circ F$ and $\mu': F \circ F' \longrightarrow \mathrm{Id}_{\mathbf{A}'}$. As usual, an equivalence in this case means that the functor is full, faithful, and essentially surjective.

We will find that, in general, the models of computation do not fit neatly into the structure of a category. Rather, they form a bicategory. This is a structure introduced by Jean Bénabou in 1967, that is weaker than a category (in the sense that it does not satisfy some of the axioms of a category) and it is a type of higher category (there are not only objects and morphisms, but also 2-cells.)

Definition 2.18. A *bicategory* or a *weak 2-category* $\mathbf{B}$ is a collection of

- Objects or 0-cells $a, b, c, \ldots$
- Between two objects, say a and b, there are morphisms or 1-cells $f: a \longrightarrow b$. The collection of all morphisms from a to b forms a category and is denoted $\mathrm{Hom}_\mathbf{B}(a, b)$.
- Between two morphsims, say f and g, there are 2-cells $f \Longrightarrow g$.

Furthermore,

- for any three objects, $a, b,$, and c of **B**, there is a composition functor

$$\circ: \text{Hom}_{\mathbf{B}}(b, c) \times \text{Hom}_{\mathbf{B}}(a, b) \longrightarrow \text{Hom}_{\mathbf{B}}(a, c); \qquad (2.13)$$

- for every b in **B**, there is an identity morphism $\text{id}_b : b \longrightarrow b$.

Rather then the usual rule for a category – that the composition is associative, and the identities act like a unit for the composition – in a bicategory, we insist that there are the following 2-cells.

- For every three composable morphisms f, g, and h in **B**, there is an "associator" isomorphic 2-cell

$$\alpha_{f,g,h}: (h \circ g) \circ f \Longrightarrow h \circ (g \circ f) \qquad (2.14)$$

which is natural in f, g, and h.
- For every morphism $f: a \longrightarrow b$ in **B** there are the following two "identity constraints" isomorphic 2-cells:

$$\iota_f: f \circ \text{id}_a \Longrightarrow f \qquad \text{and} \qquad \iota'_f: \text{id}_b \circ f \Longrightarrow f. \qquad (2.15)$$

These 2-cells must satisfy a rule like the pentagon coherence condition of a monoidal category as well as a coherence condition that shows that associators cohere with the identity constraints.

Notice that every category is a type of bicategory where all the 2-cells (including the associators and the identity constraints) are identity 2-cells. There are no non-trivial 2-cells.

One can formulate the notion of a functor between two bicategories. The functor has to respect the associator and identity constraints (perhaps up to a 2-cell). We will not need this because, for the most part, we will deal with functors from a bicategory to a category (which only has trivial 2-cells). This entails that most of the coherence conditions will easily be satisfied.

One then proceeds to discuss a *monoidal bicategory*, a *braided monoidal bicategory*, a *symmetric monoidal bicategory*, and appropriate functors between them. This brings up a giant morass of coherence conditions. Essentially, the symmetric monoidal structure has to cohere with the structure of the bicategory. Rather than getting bogged down by listing all the coherence conditions, we focus on the what the weak categories contain and point to the literature at the end of this section to learn more.

Research Project 2.19. Symmetric and braided monoidal categories and bicategories were first formulated to deal with topics in geometry and physics. It would be nice to understand why these structures arise in those fields. With this knowledge to hand, while you read this text, see how these categorical structures which consist of computable functions and the various models of computation fit in with the geometric and physical ideas. Does the geometric and physical notions have anything to say about computable functions or models of computation? Do the structures that we meet in this text have anything to contribute to our understanding of geometry or physics?

Further Reading

The notions of a slice and comma categories can be found in Section II.6 of Mac Lane (1998) and pages 3, 13, and 47 of Barr and Wells (1985). Symmetric monoidal categories can be found in Chapter XI of Mac Lane (1998) and Chapters XI–XIV of Kassel (1995). Bicategories can be found in Bénabou (1967), the original paper, Leinster (1998), and Section 9 of Street (1996). Braided monoidal bicategories and symmetric monoidal bicategories can be found in Gurski (2011), Shulman (2010), and sources therein.

3 Models of Computation

Since we are going to deal with the questions "Which functions are computable?" and "Which functions are efficiently computable?" we better first deal with the question "What is a computation?" We all have a pretty good intuition that a computation is a process that a computer performs. Computer scientists have given other, more formal, definitions of a computation. They have described different models where computations occur. These models are virtual computers that are exact and have a few simple rules. In most textbooks, one computational model is employed. As category theorists, we have to look at several models and see how they are related. This affords us a more global view of computation.

3.1 The Big Picture

We have united all the different models that we will deal with in Figure 3.1. Warning: this figure can look quite intimidating the first time your see it. Fear not dear reader! We will spend §§3.1, 3.2, 3.3, and 3.4 explaining this diagram and making it digestible. We call the diagram *The Big Picture*. It has a center and three spokes coming out of it. These will correspond to different ways of modeling computations. Most of the rest of this text will concentrate on the top spoke which is concerned with performing computations by manipulating strings.

To make *The Big Picture* less intimidating. All the categories are essentially

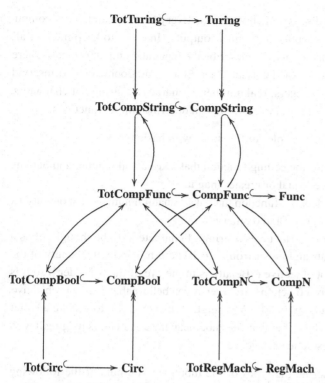

Figure 3.1 "The Big Picture" of functions and models of computation

symmetric monoidal categories. All the functors are essentially symmetric monoidal functors. All horizontal arrows are inclusion functors which are the identity on objects. Almost every category comes in two forms: (i) all the possible morphisms which include morphisms that represent partial functions, – that is functions that do not have outputs for certain inputs; and (ii) the subcategory of total morphisms, i.e., morphisms where every input has an output. The diagram has a central line consisting of different types of functions that the models of computation compute. This line will be our focus. There are three spokes pointing to that line. These correspond to three types of models of computation: (i) the top spoke corresponds to models that manipulate strings; (ii) the lower right spoke corresponds to models that manipulate natural numbers; and (iii) the lower left spoke corresponds to models that manipulate bits.

In all these categories, the composition corresponds to sequential processing (that is, performing one process after another). The monoidal structure corresponds to parallel processing. The symmetric monoidal structure corresponds to the fact that the order in parallel processing is easily interchanged.

We need a little discussion about types. Every function and every computational device takes inputs and returns outputs. In order to keep track of all the different functions we usually describe the input and output as types. There are *basic types* such as Nat, Int, Real, Char, String, and Bool which correspond to natural numbers, integers, real numbers[1], characters, strings of characters, and Booleans (0 and 1). Using these we can describe simple functions.

Example 3.1. Some examples of functions with basic types.

- $\lceil - \rceil$: Real $\longrightarrow$ Int is the ceiling function that takes a real number and outputs the smallest integer equal or greater than it.
- Prime: Nat $\longrightarrow$ Bool is a function that takes a natural number and outputs a 1 if the number is a prime and 0 otherwise.
- Beautiful: String $\longrightarrow$ Nat takes a string of characters (to be thought of as a poem) and outputs an integer from 1 to 10 determined by the beauty of the poem. This is not objective ("Beauty is in the eye of the beholder") so it is probably not really a function. And even if you believe that beauty is objective, it is probably not computable. You might get a computer to imitate a belief that most humans have, but that does not mean the computer is judging it with a comparative aesthetic experience.

Exercise 3.2. Describe the types of the inputs and outputs of the following functions.

- A function that determines the length of a string of characters.
- A function that takes a natural number and outputs a written description of that number. For example $42 \longmapsto$ "forty-two".
- A function that takes a computer program and determines how many lines of code there are in the program.

Basic types are not enough. We need operations on types to form other types.

- **List types**: Given a type T, we can form T^* which corresponds to finite lists of elements of type T. For example, Bool* is the type of lists of Booleans and Nat* corresponds to finite lists of natural numbers.
- **Function types**: Given types T_1 and T_2, we form type $\mathrm{Hom}(T_1, T_2) = T_2^{T_1}$. This type corresponds to functions that take inputs of type T_1 and output T_2 types.
- **Product types**: Given types T_1 and T_2, we form type $T_1 \times T_2$. This type will correspond to pairs of elements, the first of type T_1 and the second of type

---

[1]Finite computers with finite memory cannot deal with arbitrary large natural numbers or integers. They also do not deal with arbitrary real numbers. We will not be bothered with this issue and permit numbers of any size and precision when working with our models.

T_2. When there is a tuple of types, we will call them a *sequence of types*. For example Seq = Nat × String × Bool or Seq' = Float × Char^{Nat} × Nat* × Int. Given two sequences of types, we can concatenate them. For example the concatenation of Seq and Seq' is

$$Seq × Seq' = (Nat × String × Bool) × (Float × Char^{Nat} × Nat^* × Int)$$
$$= Nat × String × Bool × Float × Char^{Nat} × Nat^* × Int.$$

For a type T, we shorten $T × T$ as T^2 and $T × T × T$ as T^3. In general T^n will be a sequence of n lots of T. The type T^0 we take to mean * which is called the terminal type. It is used to pick out an element of another type. A function $f : * \longrightarrow$ Seq picks out an element of type Seq.

Example 3.3. Here are a few functions that use compound types:

- DayOfWeek: String × Nat × Nat $\longrightarrow$ Nat is a function that takes a date as the name of the month, the day of the month, and the year. The function returns the day of the week (1,2,...,7) for that date.
- Comp: $T_3^{T_2} × T_2^{T_1} \longrightarrow T_3^{T_1}$ is a function that corresponds to function composition. It takes an $f : T_1 \longrightarrow T_2$ and a $g : T_2 \longrightarrow T_3$ and outputs $g \circ f : T_1 \longrightarrow T_3$.
- Inverse: $T_2^{T_1} \longrightarrow T_1^{T_2}$ takes a function and returns the inverse of the function (if it exists). This will arise in our discussion of cryptography.

Exercise 3.4. Describe the following as functions from sequences of types to sequences of types.

- A function that takes a list of natural numbers and returns the maximum of the numbers, the mean of the numbers, and the minimum of the numbers.
- A function that takes (i) a function from the natural numbers to the natural numbers, and (ii) a natural number. The function should evaluate the given function on the number.
- A function that takes two natural numbers and determines if they are the last twin primes. That is, (i) both numbers are prime, (ii) they are of the form n and $n + 2$, and (iii) there are no greater twin primes.

There are much more complicated types that are needed to describe all types of functions. There are "coproduct types" and things called "dependent types." However, we will not use them in our presentation.

Advanced Topic 3.5. These types are the beginning of several areas of theoretical computer science. *Type theory* studies complicated types and how they relate to computation. It also works with higher-order logics which can function as a foundation for all of mathematics. Category theorists have been making categories of types and looking at all of their possible structures (see Asperti and Longo, 1991; Barr and Wells, 1999.) This is also related to λ-*calculus* or *lambda calculus* which is another way of describing computations (see Lambek and Scott, 1986.) The theory of types is also the beginning of a current, very hot area of research called *homotopy type theory*. This has all types of connections to foundations of mathematics and proof checkers (see Univalent, 2013.)

Now that we have the language of types down pat, let us move on to our most important definition.

Definition 3.6. The category **Func** consists of all functions. The objects are sequences of types. The morphisms in **Func** from Seq to Seq′ are all functions that have inputs from type Seq and outputs of type Seq′. We permit all functions including partial functions and functions that computers cannot compute. The identity functions are obvious. Composition in the category is simply function composition. For example, if $f: \mathsf{Seq}_1 \longrightarrow \mathsf{Seq}_2$ and $g: \mathsf{Seq}_2 \longrightarrow \mathsf{Seq}_3$ are two functions, then $g \circ f: \mathsf{Seq}_1 \longrightarrow \mathsf{Seq}_3$ is the usual composition. The category also has a monoidal structure. On objects, the monoidal structure is concatenation or product of sequences of types. Given $f: \mathsf{Seq}_1 \longrightarrow \mathsf{Seq}_2$ and $g: \mathsf{Seq}_3 \longrightarrow \mathsf{Seq}_4$, their tensor product is

$$f \otimes g: (\mathsf{Seq}_1 \times \mathsf{Seq}_3) \longrightarrow (\mathsf{Seq}_2 \times \mathsf{Seq}_4) \tag{3.1}$$

which corresponds to performing both functions in parallel. While the monoidal product is not strictly associative, thanks to general coherence theory, we will think of **Func** as a strictly associative monoidal category. The symmetric monoidal structure comes from the trivial function that swaps sequences of types, i.e., $\mathsf{tw}: \mathsf{Seq} \times \mathsf{Seq}' \longrightarrow \mathsf{Seq}' \times \mathsf{Seq}$. We leave the details to the reader.

Exercise 3.7. What is the unit of the symmetric monoidal category structure?

Exercise 3.8. Show that the function tw satisfies the axioms making **Func** into a symmetric monoidal category.

Definition 3.9. The category of all computable functions, called **CompFunc**, is a symmetric monoidal subcategory of **Func** which has the same objects. The morphisms of this subcategory are functions (including partial functions) that a computer can compute. A computer can compute a partial function if for any input for which there is an output, the computer will give that output, and if there is no output, the computer will give no output and might go into an infinite loop. In a sense, a large part of theoretical computer science is dedicated to giving an exact definition of this subcategory of **Func**.

There is a further symmetric monoidal subcategory **TotCompFunc** which contains all the total computable functions. These are functions for which every input has an output and a computer can compute the function.

Both **CompFunc** and **TotCompFunc** have symmetric monoidal category structures similar to **Func**. There are obvious symmetric monoidal inclusion functors

$$\textbf{TotCompFunc} \hookrightarrow \textbf{CompFunc} \hookrightarrow \textbf{Func} \qquad (3.2)$$

which are the identity on objects.

Example 3.10. Here are a few examples of functions and the categories to which they belong.

- $\text{Det}^n : (\text{Real}^n)^n \longrightarrow \text{Real}$ is the determinant of an $n \times n$ matrix. (Notice that for each n there is a determinant function. To describe the determinant function for all square matrices, we would need *dependent types*.) This function is total and computable.
- Sqrt: $\text{Real} \longrightarrow \text{Real}$ takes a real number and returns its positive real square root. This is a computable partial function and hence is in **CompFunc**. By the way, $\pm\sqrt{} : \text{Real} \longrightarrow \text{Real} \times \text{Real}$ takes a real number to its positive and negative real roots and is also in **CompFunc**.
- Der: $\text{Real}^{\text{Real}} \longrightarrow \text{Real}^{\text{Real}}$, the derivative function is only partially computable. The input function must be differentiable. There is also the integral function which is roughly Int: $\text{Real}^{\text{Real}} \longrightarrow \text{Real}^{\text{Real}}$ and is also in **CompFunc**. The fundamental theorem of calculus essentially says that Der is the inverse of Int, that is, Der $\circ$ Int $=$ Id $=$ Int $\circ$ Der.

Exercise 3.11. Which category do the following functions belong to?

- Matrix multiplication $* : (\text{Real}^m)^n \times (\text{Real}^n)^p \longrightarrow (\text{Real}^m)^p$.
- Empty: $\text{String} \longrightarrow \text{Bool}$. The function that determines if a program stops on any input, or always goes into an infinite loop. That is, take a program and see if the domain of the program is empty (see Example 4.12.)
- $\dot{-} : \text{Nat} \times \text{Nat} \longrightarrow \text{Nat}$. The natural number subtraction which is defined as $x \dot{-} y = x - y$ if $x \geq y$ and 0 otherwise.

In Definition 3.9 we described computable functions as functions "that a computer can compute." The obvious question is what type of computer are we discussing? What computer process is legitimate? This section will give several answers to that question.

3.2 Manipulating Strings

Let us go through the top spoke of *The Big Picture*.

Definition 3.12. The category **CompString** is the subcategory of **CompFunc** that deals with functions from strings to strings. The objects are powers of String types. This means the objects are $String^0 = *$ (which is the terminal type), $String^1 = String$, $String^2 = String \times String$, $String^3 = String \times String \times String, \ldots$. The morphisms of this category are computable functions between powers of String types. There are partial functions in this category. The symmetric monoidal structure is similar to **Func**.

The subcategory **TotCompString** has the same objects as **CompString** but with only total computable string functions. There is an obvious symmetric monoidal inclusion functor **TotCompString** $\hookrightarrow$ **CompString**.

The symmetric monoidal inclusion functor

$$\text{Inc} \colon \textbf{CompString} \hookrightarrow \textbf{CompFunc}$$

takes $String^n$ in **CompString** to the same object in **CompFunc**. Every computable string function $f \colon String^m \longrightarrow String^n$ in **CompString** goes to the same function in **CompFunc**. The functor, Inc, is injective on objects and is full and faithful. But, as we will presently see, Inc is more than just an inclusion functor.

The symmetric monoidal functor

$$F \colon \textbf{CompFunc} \longrightarrow \textbf{CompString} \tag{3.3}$$

takes an arbitrary object of **CompFunc**, e.g. Seq, to $F(\text{Seq}) = String^m$, for some m. The idea is that every sequence of types can be encoded as a power of String. The functor F will take computable function $f \colon \text{Seq} \longrightarrow \text{Seq}'$ to $F(f) \colon String^m \longrightarrow String^n$ which is a computable function that takes encoded data of the input of f to encoded output of f. In other words, F insures that all computable functions can be encoded as a computable function between powers of String.

What is the relationship between F and Inc? Since every power of String should be encoded as itself, i.e., $F(String^m) = String^m$, we have

$$F \circ \text{Inc} = \text{Id}_{\textbf{CompString}}. \tag{3.4}$$

By contrast, in general,

$$\text{Inc} \circ F \neq \text{Id}_{\textbf{CompFunc}}. \tag{3.5}$$

Even though these two functors are not equal, there is a natural transformation

$$\text{Enc} \colon \text{Id}_{\textbf{CompFunc}} \longrightarrow \text{Inc} \circ F. \tag{3.6}$$

For every sequence Seq in **CompFunc**, we have

$$\text{Enc}_{\text{Seq}} \colon \text{Seq} \longrightarrow (\text{Inc} \circ F)(\text{Seq}) \text{ or } \text{Enc}_{\text{Seq}} \colon \text{Seq} \longrightarrow String^m \tag{3.7}$$

for some m. This computable function takes an input of type Seq and outputs the encoded value of type String^m. The family of maps Enc is natural because any decent encoding must satisfy the following

$$
\begin{array}{ccc}
\text{Seq} & \xrightarrow{\text{Enc}_{\text{Seq}}} & \text{String}^n \\
f \downarrow & & \downarrow F(f) \\
\text{Seq}' & \xrightarrow[\text{Enc}_{\text{Seq}'}]{} & \text{String}^n
\end{array}
\tag{3.8}
$$

otherwise it is useless. There is also a natural transformation

$$
\text{Enc}^* : \text{Inc} \circ F \longrightarrow \text{Id}_{\textbf{CompFunc}}.
\tag{3.9}
$$

For every sequence Seq in **CompFunc**, we have

$$
\text{Enc}^*_{\text{Seq}} : \text{String}^m \longrightarrow \text{Seq}
\tag{3.10}
$$

for some m.

Technical Point 3.13. It should be pointed out that, in general, $\text{Enc}^*_{\text{Seq}}$ is not $\text{Enc}^{-1}_{\text{Seq}}$ nor is either one invertible. This can be seen by looking at Seq = Bool or even Seq = Bool^n. There is no way that the finite amount of information that can be represented in the type Bool^n can be isomorphically mapped to the potentially infinite amount of information represented in the type String^m for any m.[2] There is a similar problem in isomorphically encoding the uncountably infinite amount of information represented in Real with the countably infinite information in String^m. The fact that neither Enc nor Enc* are invertible insures that the pair Inc and F are neither equivalences nor an adjunction.

The natural transformations Enc and Enc* are very helpful in seeing how the category **CompString** sits inside **CompFunc** (see Figure 3.2.) Every function

---

[2]There is a slight asymmetry between all the types. Whereas each of the types Nat, Int, Real, and String can all represent an infinite amount of values, the type Bool can only represent two possible values: True and False. There are at least two possible ways of "leveling the playing field" with the types.

First, we can restrict each of the types Nat, Int, Real, and String to only represent a (large but) finite number of elements. This is in accord with practical real computers which can only store a large, but finite, amount of data for each data type. The problem with this solution is that it is very unnatural to have types that only represent a finite number of values. It goes against the whole grain of theoretical computer science. For example, we will see that there are an infinite number of Turing machines and each one of them will have a number.

Another possible solution is to substitute the type Bool* for the type Bool. Rather than talking about Boolean variables, talk about potentially infinite sequences of Boolean variables. In this case, every type can describe a potentially infinite amount of data. The problem with this solution is that it is natural to have the Bool type. As we will see, one of the central concerns in theoretical computer science is decision problems which are about morphisms with codomain Bool.

In this work, we choose to not "level the playing field" and work with the asymmetry of types. (I am grateful to David Spivak for pointing this issue out to me and for helping me resolve it.)

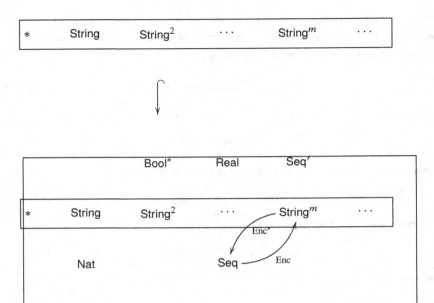

Figure 3.2 The inclusion of **CompString** into **CompFunc**.

$f\colon \mathsf{Seq} \longrightarrow \mathsf{Seq}'$ in **CompFunc** can be computed as

$$\mathsf{String}^m \xrightarrow{\ \mathrm{Enc}^*_{\mathsf{Seq}}\ } \mathsf{Seq} \xrightarrow{\ f\ } \mathsf{Seq}' \xrightarrow{\ \mathrm{Enc}_{\mathsf{Seq}'}\ } \mathsf{String}^n$$

in **CompString**.

Let us return to the Inc functor, and see what we can say about it. As is common knowledge, a functor $J\colon \mathbf{A} \longrightarrow \mathbf{B}$ is essentially surjective if for all b in $\mathbf{B}$ there is an a in $\mathbf{A}$ such that there is a natural isomorphism $J(a) \longrightarrow b$. Keep this in mind, as the next definition weakens this notion.

Definition 3.14. A $J\colon \mathbf{A} \longrightarrow \mathbf{B}$ is an *inherently surjective functor* if for all b in $\mathbf{B}$ there is an a in $\mathbf{A}$ such that there exist two maps

$$e\colon J(a) \longrightarrow b \qquad \text{and} \qquad e^*\colon b \longrightarrow J(a) \qquad (3.11)$$

natural in b.

Notice that if $e^* = e^{-1}$ then J is an essentially surjective functor.

Let $J = \mathrm{Inc}$ and for every Seq in **CompFunc**, let $F(\mathsf{Seq})$ be the object in **CompString** that has the two morphisms e and e^*. We have proved the following theorem:

Theorem 3.15. *The inclusion functor* Inc: **CompString** $\longhookrightarrow$ **CompFunc** *is injective on objects, full, faithful, and inherently surjective. The same is also true for the inclusion functor* **TotCompString** $\longhookrightarrow$ **TotCompFunc**.

Let us continue up the top spoke of The Big Picture. In the 1930s, Alan Turing wondered about the formal definition of a computation. He came up with a model we now call a *Turing machine* which manipulates strings. Turing based his work on the idea that mathematicians do computations. They manipulate the symbols of mathematics in different ways when they are in different states. For example, if a mathematician sees the statement $x \times (y + z)$ and is in the distributive state, she will then cross out that statement and write $(x \times y) + (x \times z)$. In the same way, a Turing machine has a finite set of states that describe what actions the machine should perform. Just as a mathematician writes her calculation on a piece of paper, so too, a Turing machine performs its calculations on paper. Turing was inspired by ticker tape machines and typewriter ribbons to define his paper as a thin tape that can only have one character per space at a time. In this text, our version of the machine has several tapes that are used for input, several tapes for working out calculations, and several tapes for output. For every tape, there will be an arm of the Turing machine that will be able to see what is on the tape, change one symbol for another, and move to the right or the left of that symbol. We envision the Turing machine as in Figure 3.3. A typical rule of a Turing machine will say something like "If the machine is in state q_{32} and it sees symbol x_1 on the first tape and x_2 on the second tape, . . . , and the symbol x_t in the tth tape, then change to state q_{51}, make the symbol in the first tape to a y_1 and the symbol in the second tape to y_2, . . . the symbol in the tth tape into a y_t, also move to the left in the first tape, the right in the second tape, . . . , the right in the tth tape." In symbols we can write this as

$$\delta(q_{32}; x_1, x_2, \dots, x_t) = (q_{51}; y_1, y_2, \dots, y_t; L, R, R, L, \dots, R); \qquad (3.12)$$

here δ describes the rules or the program of the Turing machine. There will be many such rules which forms a function δ that is called the *transition function* of the Turing machine. For a Turing machine with a set of states Q, an alphabet Σ, and t tapes, the transition function is defined as

$$\delta \colon Q \times \Sigma^t \longrightarrow Q \times \Sigma^t \times \{L, R\}^t. \qquad (3.13)$$

There is an obvious question that we left out: how long should the paper be? Turing realized that if you limit the size of the paper, then you will only be able to compute certain less complicated functions. Since Turing was only interested in whether or not a function was computable, and not whether or not it it was computable with a certain sized paper, therefore Turing insisted that the

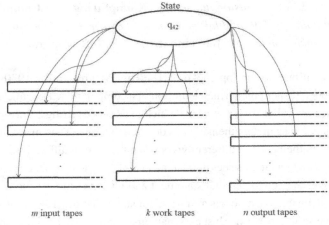

m input tapes *k* work tapes *n* output tapes

Figure 3.3 A Turing Machine

paper be infinitely long, with no bound on how much room there would be for calculations. It was only thirty years later that theoretical computer scientists started being concerned with how much space resources would be required to compute certain functions. (We will see more of this in §5 while learning complexity theory.)

A computation occurs when data is put on the input tapes and the Turing machine is in a special starting state. The machine then follows the rules on how to manipulate the input strings, computes on the work tapes, and writes the results on the output tape. It will go through many states manipulating the strings. There are two possible outcomes that can happen with this process: (i) the Turing machine can come to an accepting state with symbols on its tapes for which there is no further rule. The machine then halts. Or (ii) the Turing machine continues forever in an infinite loop.

Example 3.16. Here are three Turing machines. Rather than describing the Turing machine by giving each rule, we shall provide a "higher-level" description that shows how the Turing machine performs the task.

- **Addition.** $+\colon \mathsf{Nat} \times \mathsf{Nat} \longrightarrow \mathsf{Nat}$. The input numbers are given as strings of digits on two input tapes. The Turing machine will go to the right side of both input strings and work its way to the left. At the same time, the Turing machine will move on the output tape one space more to the right than the largest input. One space of a single work tape will either have a 1 or a 0 for a carry digit. Once in place, the computer will add the numbers one digit at a time adding the necessary carry digit. The result will go on the output tape.
- **Multiplication.** $\cdot\colon \mathsf{Nat} \times \mathsf{Nat} \longrightarrow \mathsf{Nat}$. This is a little harder than the addition.

The computer starts at the right side of both input strings. Assume that the strings are of length m and n. The output will start in position $m + n$. The machine will multiply the appropriate digits of one number with the other number while keeping a running sum of the products. At the end, the output is put in its correct position.

- **Sorting**.   Sort: $\text{Nat}^n \longrightarrow \text{Nat}^n$ sorts an n-tuple of numbers into non-descending order. The Turing machine goes through all n of the input tapes and determines the largest number. When it is found, it changes that input tape to 0 and places that number on the nth output tape. Repeat this and place the largest on the $(n - 1)$th output tape. Do this n times.

Exercise 3.17. Give descriptions for Turing machines that perform the following functions: (i) Determine if a word in alphabet $\{a, b\}$ is of the form $a^n b^n$. (ii) Determine if a number is prime. (iii) Exponentiation.

We do not want to look at only one Turing machine by itself. As category theorists, we want to look at the collection of *all* Turing machines and how they interact. We are going to form a symmetric monoidal bicategory called **Turing** where the objects will be the natural numbers. A Turing machine with m input tapes and n output tapes will then correspond to a morphism $T: m \longrightarrow n$ in the bicategory. The category $\text{Hom}_{\textbf{Turing}}(m, n)$ consists of all Turing machines with m input tapes and n output tapes. The symmetric monoidal bicategory will have isomorphic 2-cells between two Turing machines which we draw as follows:

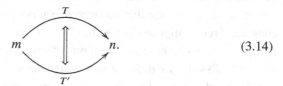

$$\tag{3.14}$$

These 2-cells will be generated in a way that will be explained shortly. The isomorphic 2-cells will only go between two Turing machines that are "essentially the same" and that perform the same function.

Turing machines compose in two ways. One is sequential composition (i.e., perform one Turing machine's operations followed by another). The other is parallel composition (i.e., perform the operations of the two Turing machines at the same time.) Let us look at both of these operations carefully.

Sequential composition of Turing machines is as follows. If $T: m \longrightarrow n$ and $T': n \longrightarrow p$ are Turing machines, then $T' \circ T: m \longrightarrow n \longrightarrow p$ is a Turing machine. In order to properly compose them, we must first make sure that they do not have any common states. If T' has common states with T, rename them. Once this is done, we can compose the machines by making the output tapes of

T become the input tapes of T'. After the T machine halts, the machine will go to the start state of the T' machine. If the T machine does not halt, the T' machine never even begins. The computation is accepted or rejected when the computation reaches the accepting or rejecting state of T'. For every n, there is an identity Turing machine $T_{id} \colon n \longrightarrow n$ which takes all the data on the input tapes to the output tapes. When any Turing machine is composed with the identity Turing machine, the same function is performed as the original Turing machine.

Parallel composition of Turing machines happen as follows. If $T \colon m \longrightarrow n$ and $T' \colon m' \longrightarrow n'$ are Turing machines, then $T \otimes T' \colon m + m' \longrightarrow n + n'$ is their parallel composition. If T has k work tapes and T' has k' work tapes, then $T \otimes T'$ has $k + k'$ work tapes. The Turing machine $T \otimes T'$ has to be defined by doing both processes at the same time. The set of states is the product of the two sets of states. At each point in time, this Turing machine does what both machines would do. We leave the details for an exercise. There is a trivial Turing machine $T_* \colon 0 \longrightarrow 0$ that has no input or output. When you parallel compose a Turing machine with the trivial Turing machine you get the same function as the function of the original Turing machine.

Exercise 3.18. Let Turing machine T_1 have transition function δ_1 and Turing machine T_2 have transition function δ_2. Describe the transition function of $T_1 \otimes T_2$ in terms of δ_1 and δ_2.

Technical Point 3.19. Neither sequential composition nor parallel composition are associative. To see that sequential composition is not associative is to consider three contiguous Turing machines $T \colon m \longrightarrow n$ and $T' \colon n \longrightarrow p$ and $T'' \colon p \longrightarrow q$ and we have a uniform way of changing common state names, then $T'' \circ (T' \circ T)$ will have different states names than $(T'' \circ T') \circ T$. It is important to point out that these two different Turing machines are "essentially the same" and they will perform the same computable function. The problem with parallel composition simply arises from the fact that the set of states $Q_1 \times (Q_2 \times Q_3)$ is not the same as the set of states $(Q_1 \times Q_2) \times Q_3$. Again, even though they are different Turing machines they are "essentially the same" and they compute the same computable function.

Another problem is that the identity Turing machine does not act like a unit for the sequential composition operation, and the trivial Turing machine does not act like a unit for the parallel composition operation. In detail, the identity Turing machine $Id \colon m \longrightarrow m$ composed with Turing machine $T \colon m \longrightarrow n$ is $T \circ Id \colon m \longrightarrow m \longrightarrow n$. While this Turing machine performs the same computable function and is "essentially the same" as $T \colon m \longrightarrow n$, the two are nevertheless not the same Turing machine. We leave it to the reader to show that the trivial Turing machine does not act like a unit for the parallel composition.

Because of the associativity and unit problems, the collection **Turing** does not form a category. This is why we need the notion of a monoidal bicategory.[3] Such structures are almost categories: they do not satisfy the axioms of a category, but they do contain iso 2-cells. In our case, the presence of iso 2-cells is a way of saying that the morphisms are "essentially the same."

In detail, although the composition of three Turing machines is not associative, there is an iso 2-cell between them. Explicitly, if there are three contiguous Turing machines $T: m \longrightarrow n$ and $T': n \longrightarrow p$ and $T'': p \longrightarrow q$, then there is the following 2-cell:

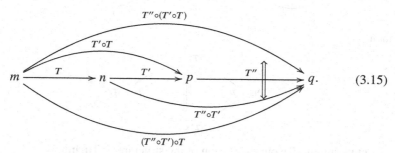

$$ (3.15) $$

Saying that the identity Turing machines act as a unit to composition means there exist the following two iso 2-cells:

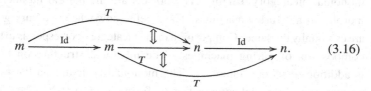

$$ (3.16) $$

All these iso 2-cells must satisfy coherence rules. These 2-cells must respect the sequential composition of morphisms. This means that if

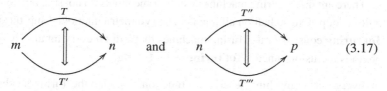

$$ (3.17) $$

---

[3]Another way to deal with this problem is to talk about equivalence classes of Turing machines, i.e., although $T \circ \mathrm{Id} \neq T$ we insist that $T \circ \mathrm{Id} \sim T$. In Yanofsky (2011) this method is used to deal with the problem. Different equivalence relations are discussed in Yanofsky (2017).

then

$$(3.18)$$

Furthermore, the 2-cells have to respect the tensor product of morphisms. This means that if

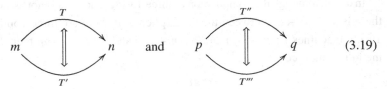

$$(3.19)$$

then

$$(3.20)$$

Let us summarize our discussion of the structure of **Turing**.

Definition 3.20. The collection of all Turing machines forms a symmetric monoidal bicategory **Turing**. The objects are the natural numbers and the morphisms are Turing machines. The 2-cells show when two Turing machines are essentially the same. Composition in the bicategory corresponds to sequential composition of Turing machines. The monoidal structure on the objects is addition of natural numbers. The monoidal structure on the morphisms corresponds to parallel composition of Turing machines. It is easy to see that one can go from Turing machine $T \otimes T'$ to Turing machine $T' \otimes T$ and hence **Turing** has a symmetric structure.

There are many Turing machines where some inputs do not stop but go into an infinite loop. Others halt on all inputs. The symmetric monoidal sub-bicategory **TotTuring** consists of the Turing machines that halt on every input. There is an obvious inclusion functor **TotTuring** $\hookrightarrow$ **Turing**.

Every Turing machine describes a function. We map the Turing machines to the functions they compute. This map is a functor **Turing** $\longrightarrow$ **CompString** from the symmetric monoidal bicategory **Turing** to the symmetric monoidal category **CompString** that takes object m to String^m. A Turing machine with m inputs and n outputs will go to its function from String^m to String^n. The iso

2-cells in **Turing** will go to identity morphisms in **CompString**. The functor is bijective on objects but is not the identity on objects (n goes to String^n not n).

It is believed that one can go the other way. Given any computable function, we can find a Turing machine that computes it. This is the content of the *Church–Turing thesis* which says that any computable function can be computed by a Turing machine. In our categorical language this means that the functor **Turing** $\longrightarrow$ **CompString** is full. This is called a "thesis" rather than a "theorem" because it has not been proven and probably cannot be proven. The reason why is that there is no perfect definition of what it means to be computable function. How can we prove that every computable function can be computed by a Turing machine when we cannot give an exact characterization of what we mean by computable function? In fact, some people define a computable function as a function which can be computed by a Turing machine. If we take that as a definition, then the Church–Turing thesis is true but has absolutely no content. Be that as it may, most people take the Church–Turing thesis to be true. Turing machines have been around since the 1930s and no one has found a computable function that a Turing machine cannot compute. In the next section we will describe functions that cannot be computed by a Turing machine and hence cannot be computed by any computer.

Exercise 3.21. The functor **Turing** $\longrightarrow$ **CompString** is full but it is not faithful. Why is it not faithful?

The category **Turing** describes every possible computational procedure. There is, however, a small proper subcategory of **Turing**, namely **Turing**(1, 1), that also describes every possible computational procedure. This subcategory can be thought of as the one-object category with all the morphisms from 1 to 1 in **Turing**. This contains all Turing machines with one input tape and one output tape. There is an obvious inclusion of **Turing**(1, 1) into **Turing**. However, the main point is that every Turing machine is equivalent to some Turing machine in **Turing**(1, 1), i.e., there is a functor $F \colon$ **Turing** $\longrightarrow$ **Turing**(1, 1) such that the following two triangles commute:

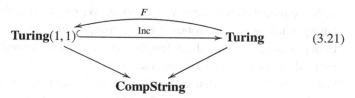

$$\text{(3.21)}$$

where $F \circ \text{Inc} = \text{Id}$ but, in general, $\text{Inc} \circ F \neq \text{Id}$. A multi-tape Turing machine is computed by a Turing machine in **Turing**(1, 1) in the following manner. If there are, say, three tapes with contents

$$\boxed{\text{SRTHGKMG}} \quad , \quad \boxed{52682406} \quad , \quad \text{and} \quad \boxed{\text{AnAngoAN}} ,$$

then the single tape of the computing machine will look like this:

$$\boxed{\text{SRTHGKMG\#52682406\#AnAngoAN}} .$$

The contents of the tape are separated by the new symbol #. The computing machine with a single arm will then go back and forth and perform what it is supposed to do on each tape. While it is very inefficient, the computing machine can nevertheless perform the same operations.

It should be stressed that **Turing**$(1, 1)$ is not a monoidal category and the above two functors are not monoidal functors. Most books go even further and use Turing machines that have a single one-way infinite tape. This single tape serves as input tape, work tape, and output tape. The above trick works in this case as well.

Before we close our discussion of Turing machines, let us mention a few points.

- A warning about Turing machines is in order. No one programs with a Turing machine. They are just a pain in the neck and they are a hundred times more unpleasant than programming in assembly language. The trick is to accept the Church–Turing thesis and instead of thinking of Turing machines, think of any computer. If a computer can perform a task, then so can a Turing machine. In a sense, Turing machines and all the other models that we will meet, are red herrings. They are distractions that keep you from the important invariant idea: computable functions. How they are computed is irrelevant. This is the content of the famous dictum (wrongly) attributed to Edsger Dijkstra: "Computer Science is no more about computers than astronomy is about telescopes."

- There are many different definitions of a Turing machine. Most have one tape. Some have tapes that are infinite in two directions. Some insist that when a Turing machine stops it clears its work on the tapes. Some permit the Turing machine to stay rather than to move, etc. It turns out that all these differences do not change the power of a Turing machine. What can be done with one model, can be done with another. We say that the definition of a Turing machine is very "robust." Small changes do not make a difference. Hence, although we have been less than exact with our conventions, we are justified because it really does not matter.

- What we have actually defined here is a *deterministic* Turing machine. In §5, we will meet souped-up versions of Turing machines called *nondeterministic* Turing machines.

Advanced Topic 3.22. In 1956, Claude E. Shannon, the famous founder of information theory, proved the following two interesting theorems. Every Turing machine is equivalent to a Turing machine with only two symbols in its alphabet. (This is done by adding many more states to the Turing machine.) This is to be expected, after all, our electronic computers only use two symbols: "0" and "1." In the same paper, he also proved a really shocking theorem. Every Turing machine is equivalent to a Turing machine with only two states. (This is done by adding more symbols to the Turing machine.) See Shannon (1956).

We will now discuss the bottom two spokes of *The Big Picture* in Figure 3.1. We note that most of the rest of this text will be developed with the categories and functors from the top spoke alone. The ideas and theorems of theoretical computer science can be stated using the language of any of the spokes. We chose to focus on Turing machines for historical reasons. If you are not interested in other models of computation, you can skip them. We include them because we want to highlight the diversity of computational models and how category theory unifies the ideas.

3.3 Manipulating Natural Numbers

While Turing thought of a computation as manipulating strings, others such as Alonzo Church, Kurt Gödel, Stephen Kleene, Joachim Lambek, Marvin Minsky, John C. Shepherdson, and Hao Wang thought of a computation as something to do with manipulating natural numbers. They dealt with functions whose input and output were sequences of natural numbers.

This brings us to define the following categories.

Definition 3.23. The objects of the category **CompN** are types $Nat^0 = *$, $Nat^1 = Nat$, $Nat^2 = Nat \times Nat$, $Nat^3 = Nat \times Nat \times Nat, \ldots$. The morphisms are all computable functions from powers of Nat to powers of Nat (including partial functions). The symmetric monoidal category structure is the same as that of **Func**.

There is a subcategory **TotCompN** with the same objects, but with only total computable functions. There is an obvious inclusion **TotCompN** $\hookrightarrow$ **CompN**.

There is an inclusion Inc from **CompN** into **CompFunc** that takes Nat^m to Nat^m. Just as we can encode any sequence of types as powers of strings, so too we can encode any sequence of types as powers of natural numbers. From this we get the analog of our result in Theorem 3.15.

Theorem 3.24. *The inclusion functor* Inc: **CompN** $\hookrightarrow$ **CompFunc** *is injective on objects, full, faithful, and inherently surjective. The same is also true for the inclusion functor* **TotCompN** $\hookrightarrow$ **TotCompFunc**.

Just as Turing machines are methods for manipulating strings, *register machines* are methods for manipulating natural numbers. These machines are basically programs in a very simple programming language where variables can only hold natural numbers. These programs use three different types of variables, namely: $X_1, X_2, X_3, \ldots$ called "input variables;" $Y_1, Y_2, Y_3, \ldots$ called "output variables;" and $W_1, W_2, W_3, \ldots$ called "work variables." The register machine usually starts with the input variables initialized to the inputs and all the other variables initialized to 0. Register machines employ only the following types of operations on any variable Z:

$$ Z = Z + 1 \qquad Z = Z - 1 \qquad \text{If } Z \neq 0 \text{ goto } L, \qquad (3.22) $$

where L is some line number. These operations are called "increment," "decrement," and "conditional branch." A program is a list of such statements for various variables. The machine then follows the program. Each variable is to be thought of as a computer register that holds a natural number. The values in the output variables at the end of an execution are the output of the function. There exist certain register machines for which some of the input causes the machine to go into an infinite loop and have no output values. Other register machines halt for any input.

It should be pointed out that there are, in fact, many different types of register machines with different operations and different conventions. Even though they are different, they are all equivalent and can program any computable function. Just like there are many definitions of Turing machines, register machines are also robust.

The programming language looks a little primitive. It does not seem that these three pithy operations could perform many functions. But this is not true. One can perform any computable function with a register machine. We begin with a few simple functions and the programs that implement them. Once we describe a program, we will use it as a module in future programs.

Example 3.25. Some programs for register machines can be seen in Figure 3.4.

Exercise 3.26. Describe register machines that perform each of the following functions. (i) If $X_1 = 0$ goto L. (ii) $Y_1 = 1$ if $Z_1 < Z_2$ else $Y_1 = 0$. (iii) $Y_1 = \lfloor \sqrt{Z_1} \rfloor$.

As category theorists we have to look at the collection of all register machines and how they interact. This collection will form a symmetric monoidal bicategory called **RegMach**. The objects will be natural numbers and a register machine with m input variables and n output variables will be denoted as $R \colon m \longrightarrow n$.

(i) `Goto L`	(ii) $Z_1 = Z_2$
1. $W_{28} = W_{28} + 1$ 2. `If` $W_{28} \neq 0$ `goto L`	1. `If` $Z_2 \neq 0$ `goto 3` 2. `Goto 6` 3. $Z_2 = Z_2 - 1$ 4. $Z_1 = Z_1 + 1$ 5. `Goto 1` 6. `Stop.`
(iii) $Y_1 = Z_1 + Z_2$	(iv) $Y_1 = Z_1 * Z_2$
1. $Y_1 = Z_1$ 2. `If` $Z_2 \neq 0$ `goto 4` 3. `Goto 7` 4. $Y_1 = Y_1 + 1$ 5. $Z_2 = Z_2 - 1$ 6. `Goto 2` 7. `Stop.`	1. $Y_1 = Z_1$ 2. `If` $Z_2 \neq 0$ `goto 4` 3. `Goto 7` 4. $Y_1 = Y_1 + Z_1$ 5. $Z_2 = Z_2 - 1$ 6. `Goto 2` 7. `Stop.`
(v) $Y_1 = Z_1 - Z_2$ if $Z_1 \geq Z_2$	(vi) $Y_1 = 1$ if X_1 is even else $Y_1 = 0$
1. $Y_1 = Z_1$ 2. `If` $Z_2 \neq 0$ `goto 4` 3. `Goto 7` 4. $Y_1 = Y_1 - 1$ 5. $Z_2 = Z_2 - 1$ 6. `Goto 2` 7. `Stop.`	1. $Y_1 = 1$ 2. `If` $X_1 \neq 0$ `goto 4` 3. `Goto 10` 4. $X_1 = X_1 - 1$ 5. `If` $X_1 \neq 0$ `goto 7` 6. `Goto 9` 7. $X_1 = X_1 - 1$ 8. `Goto 2` 9. $Y_1 = 0$ 10. `Stop.`

Figure 3.4 Examples of register machines.

The symmetric monoidal bicategory will have isomorphic 2-cells between two

register machines which we draw as

$$m \quad\quad n. \quad\quad\quad (3.23)$$

As with Turing machines there are two types of composition of register machines: sequential and parallel.

Sequential composition is not hard to define: basically the variables in the second program are changed so that they do not have any variables in common with the first program and then the second program is attached onto the end of the first program. Output variables of the first program must be set equal to input variables of the second program. For every m, there is an identity register machine $\mathrm{Id}_m : m \longrightarrow m$ that does nothing more then set all its output variables to its input variables, i.e., nothing is changed. When any register machine is sequentially composed with the identity register machine, the result is a machine that computes the same computable function as the original register machine.

Parallel composition is also not hard to formulate. One program demands m input variables and n output variables, and a second program demands m' input variables and n' output variables. Then the parallel composition of those two register machines will demand $m + m'$ input variables and $n + n'$ output variables. Before we parallel compose them, we must change the names of the any overlapping variables. The composed program will do both programs in parallel (the concept is similar to "multithreading"). Line numbers must also be adjusted. All this can be formalized with a little thought. There is also a trivial register machine with no input variables and no output variables. In fact, it need not have any work variables. This register machine does nothing. When any register machine is parallel composed with the trivial register machine, the result is a machine that computes the same computable function as the original register machine.

Similar to what we saw about Turing machines in Technical Point 3.19, we have the following.

Technical Point 3.27. Neither sequential composition nor parallel composition of register machines are associative. The way to see that sequential composition is not associative is to realise that if there is a uniform way of changing common variable names, then the two different compositions of three register machines will not have the same names of variables. It is important to point out that these two different register machines are "essentially the same" and they will perform the same computable function.

The problem with parallel composition is similarly about changing the names of the variables. The disjoint union of variable names is not associative. Again, even though they are different register machines they are "essentially the same" and they compute the same computable function.

Another problem is that the identity register machines do not act like a unit for the sequential composition operation, and the trivial Turing machine does not act like a unit for the parallel composition operation. Although the composed register machine compute the same function, the composed register machine is not exactly the same as the original register machine. The problems are similar to the ones mentioned with Turing machines.

Since there are problems with the associativity and the units of the compositions, the collection **RegMach** will not form a category. It does form a symmetric monoidal bicategory. The iso 2-cells between two register machines will be generated from associativity and units just as they were in Turing machines.

Let us summarize the structure we have for the collection all register machines.

Definition 3.28. The symmetric monoidal bicategory **RegMach** contains all register machines. The objects of the bicategory are the natural numbers. The morphisms from m to n are all register machines with m input variables and n output variables. There will be isomorphic 2-cells between register machines which perform the same operation but are different because of associativity problems or unit problems.

There is a symmetric submonoidal sub-bicategory **TotRegMach** whose objects are also the natural numbers and whose morphisms are total register machines, i.e., register machines that have values for all inputs. There is an obvious inclusion **TotRegMach** $\hookrightarrow$ **RegMach**.

There is a functor from the symmetric monoidal bicategory **RegMach** of register machines to the symmetric monoidal category **CompN**,

$$\textbf{RegMach} \longrightarrow \textbf{CompN}, \tag{3.24}$$

that on objects takes n to Nat^n and a register machine to the function it describes. 2-cells in **RegMach** go to the identity 2-cells in **CompN**, meaning that "essentially the same" register machines (as defined above) go to the same computable function. The belief that every computable function on natural numbers can be computed by a register machine means that this functor is full. This is simply another restatement of the Church–Turing thesis that we saw earlier. There is a similar full functor

$$\textbf{TotRegMach} \longrightarrow \textbf{TotCompN}. \tag{3.25}$$

In addition to register machines, there is another way of describing the morphisms in the category **CompN**. The morphisms can be generated from special morphisms using particular types of generating operations. The special morphisms in the category **CompN** are called *basic functions*:

- The **zero function** z: Nat $\longrightarrow$ Nat which is defined for all n as $z(n) = 0$.
- The **successor function** s: Nat $\longrightarrow$ Nat which is defined for all n as $s(n) = n + 1$.
- The **projection functions** for each n and for each $1 \le i \le n$, there is π_i^n: Nat^n $\longrightarrow$ Nat which is defined as $\pi_i^n(x_1, x_2, \ldots, x_n) = x_i$.

These morphisms are clearly computable and hence in **CompN**.

Exercise 3.29. Show that the basic functions are computable by describing a register machine that computes them.

There are three generating operations on morphisms in **CompN**:

- The *composition operation*: given f_1: Nat^m $\longrightarrow$ Nat, ..., f_n: Nat^m $\longrightarrow$ Nat and g: Nat^n $\longrightarrow$ Nat, there is a function h: Nat^m $\longrightarrow$ Nat defined as

$$h(x_1, x_2, \ldots, x_m) = g(f_1(x_1, \ldots, x_m), \ldots, f_n(x_1, \ldots, x_m)).$$

- The *recursion operation*: given f: Nat^m $\longrightarrow$ Nat and g: Nat^{m+1} $\longrightarrow$ Nat there is a function h: Nat^{m+1} $\longrightarrow$ Nat defined as

$$h(x_1, x_2, \ldots, x_m, 0) = f(x_1, x_2, \ldots, x_m) \tag{3.26}$$
$$h(x_1, x_2, \ldots, x_m, n + 1) = g(x_1, x_2, \ldots, x_m, h(x_1, x_2, \ldots, x_m, n)). \tag{3.27}$$

- The *μ-minimization operation*: given f: Nat^{m+1} $\longrightarrow$ Nat there is a function h: Nat^m $\longrightarrow$ Nat that is defined as follows:

$$h(x_1, x_2, \ldots, x_m) = \text{the smallest number } y \text{ with } f(x_1, x_2, \ldots, x_m, y) = 0$$
$$= \mu_y[f(x_1, x_2, \ldots, x_m, y) = 0].$$

If no such y exists, h will be undefined for those inputs.

Notice that all these generating operations produce functions that output Nat. In order to produce functions that output Nat^n, one would need to use the tensor product in the category.

Exercise 3.30. Show that if the input functions to these operations can be computed by register machines, then the output functions can also be computed by register machines. In other words, the property of being computed by a register machine is closed under composition, recursion, and μ-minimization.

One can generate morphisms in **CompN** as follows. Start with the basic functions and then perform these three operations on them. Add the resulting morphisms to the set of morphisms to perform more operations. Continue generating morphisms in this manner.

Definition 3.31. A function constructed from the basic functions using composition and recursion is called a *primitive recursive function*. A function constructed from the basic functions using composition, recursion, and μ-minimization is called a *recursive function*.

Example 3.32. Here are some primitive recursive functions.

(i) Addition	(ii) Multiplication	(iii) Exponentiation
$x + 0 = x$	$x * 0 = 0$	$x^0 = 1$
$x + (s(y)) = s(x + y)$	$x * (s(y)) = x * y + x$	$x^{s(y)} = (x^y) * x$
(iv) Predecessor	(v) Subtraction	(vi) Factorial
$P(0) = 0$	$x - 0 = x$	$0! = 1$
$P(s(x)) = x$	$x - (s(y)) = P(x - y)$	$(s(x))! = s(x) * (x!)$

Exercise 3.33. Show that the following functions are primitive recursive. (i) The sign of a number sg: Nat $\longrightarrow$ Bool, which is defined as $\mathrm{sg}(x) = 1$ if and only if $x \neq 0$. (ii) The distance between two numbers $| - |$: Nat $\times$ Nat $\longrightarrow$ Nat. (iii) The remainder when two numbers are divided rem(,): Nat $\times$ Nat $\longrightarrow$ Nat which is defined as $\mathrm{rem}(x, y) = r$ if and only if there is a q such that $y = x * q + r$. In order to make this a total function, we define $\mathrm{rem}(0, y) = 0$.

Example 3.34. Let us look at an example of a recursive function. Let $p(x)$ be a polynomial with integer coefficients. Let f: Nat $\longrightarrow$ Nat be defined as $f(m) = n$ if n is the least root of $p(x) - m$ and is undefined if no such root exists. That is, n is the smallest number such that $p(n) = m$. The function f is defined as $f(m) = \mu_n[p(n) - m = 0]$.

Example 3.35. The predicate $(? =?)$: Nat$\times$Nat $\longrightarrow$ Bool that takes two numbers and returns 1 if and only if they are equal is primitive recursive. This is because $(x = y)$ can be defined as $1 - \mathrm{sg}(|x - y|)$.

Exercise 3.36. Prove that if f: Nat $\longrightarrow$ Nat is a total injective computable function, then f^{-1} is computable.

It should be noted that there usually is more than one way to describe a primitive recursive function. This is similar to the fact that there are many Turing machines (or register machines) to describe the same function. Similar statements can be made about recursive functions.

Advanced Topic 3.37. In Chapter IX of Manin (2010), it is shown that one can describe the formation of primitive recursive functions with categorical structures called "operads" and "PROPs." Such structures arise in homotopy theory and theoretical physics.

Let us formulate categories of these functions.

Definition 3.38. The category **PRCompN** has powers of Nat as objects and the morphisms will be functions that can be described as primitive recursive functions. The category **Recursive** will have powers of Nat as objects and the morphisms will be functions that can be described as recursive functions. Obviously **PRCompN** ⟶ **Recursive** where the inclusion is the identity on objects.

We must emphasize that the morphisms in these categories are simply special morphisms in **Func**. Primitive recursive functions and recursive functions are just ways of describing (computable) functions.

It is interesting to examine which of these morphisms are in **TotCompN**. All the basic functions are in **TotCompN**. Notice that if the f_i and g of the composition operation are in **TotCompN** then so is h, i.e., **TotCompN** is closed under the composition operation. **TotCompN** is also closed under the recursion operation. This means that **PRCompN** is included in **TotCompN**. By contrast, **TotCompN** is not closed under the μ-minimization operation. That is, there could be a total f and a $x_1, x_2, \ldots, x_m$ such that there does not exist a y with $f(x_1, x_2, \ldots, x_m, y) = 0$. In that case $h(x_1, x_2, \ldots, x_m)$ is not defined. That is, h is a partial function and hence a morphism in **CompN** but not in **TotCompN**.

Theorem 3.39. Recursive = **CompN**, *i.e., recursive functions are computable functions and vice versa.*

Proof. Both categories have the same objects (the natural numbers.) By the Church–Turing thesis, every computable function can be described by a register machine. Essentially what was done in Exercises 3.29 and 3.30 was to show that every recursive function can be described by register machines. Going the other way is a little more complicated. We need to show that every register machine can be computed by recursive functions. We do not give the proof here but refer to Chapter 3 of Davis et al. (1994b), Chapter 3 of Cutland (1980), and Chapter 6 of Boolos et al. (2007). □

Summing up, we have

$$\textbf{PRCompN} \longrightarrow \textbf{TotCompN} \longrightarrow \begin{matrix}\textbf{CompN} =\\ \textbf{Recursive}\end{matrix} \longrightarrow \textbf{Func}. \tag{3.28}$$

All the functors are faithful but not full.

Advanced Topic 3.40. The morphisms in **CompN** are generated from the basic functions using the composition, recursion, and μ-minimization operations. The *Kleene normal form theorem* says that every morphism in **CompN** – that is, every computable function – can be described using *at most one* μ-minimization operation. See Theorem 3.3 of Davis et al. (1994b), Theorem X of Chapter 1 of Rogers (1987) and Theorem 3.3 of Chapter 1 of Soare (1987).

There is a beautiful categorical way of describing the category of **PRCompN** due to Román (1989), that is worthy of contemplation.

- Obviously, **PRCompN** is a category with composition.
- As we know, **PRCompN** has a monoidal category structure which is really a Cartesian product. Let us be explicit about it. The objects are powers of Nat. So $\text{Nat}^m \times \text{Nat}^n = \text{Nat}^{m+n}$. And if $f : \text{Nat}^m \longrightarrow \text{Nat}^n$ and $g : \text{Nat}^{m'} \longrightarrow \text{Nat}^{n'}$ then $f \times g : \text{Nat}^{m+m'} \longrightarrow \text{Nat}^{n+n'}$.
- There is a way of generating new elements using recursion. For a category theorist, the ability to describe morphisms via recursion means there is a *natural number object* in the category. In order to define this, let us step back a little and talk about a simple version of recursion. Given a number f in $\mathbb{N}$ and a function $g : \mathbb{N} \longrightarrow \mathbb{N}$, one can describe a function $h : \mathbb{N} \longrightarrow \mathbb{N}$ such that

$$h(0) = f, \tag{3.29}$$
$$h(s(n)) = g(h(n)). \tag{3.30}$$

This can be described in a category as a natural number object which is a diagram

$$* \xrightarrow{\quad 0 \quad} \mathbb{N} \xrightarrow{\quad s \quad} \mathbb{N}, \tag{3.31}$$

where 0 picks out the zero of the natural numbers and s is the successor function, such that for any $f \in \mathbb{N}$ and $g : \mathbb{N} \longrightarrow \mathbb{N}$, there exists an $h : \mathbb{N} \longrightarrow \mathbb{N}$ such that the following diagram commutes:

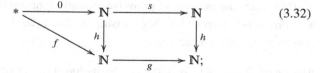

$$\tag{3.32}$$

see e.g. Barr and Wells (1985, 1999); Mac Lane (1998). Saying that the above diagram commutes is the same as saying that h is defined by the simple recursion scheme.

We can use the same scheme to generalize to any object in the category. For any object A, and any maps $f: * \longrightarrow A$, $g: A \longrightarrow A$, there is an $h: \mathbb{N} \longrightarrow A$ which makes the following commut:e

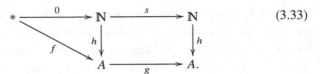

$$(3.33)$$

We need to discuss functions that have more than one input and hence need to consider a souped-up natural number object called a *parameterized natural number object*. That is, we need two objects in the category called A and B, and for every $f: A \longrightarrow B$ and $g: A \times B \longrightarrow B$ there exists an $h: A \times \mathbb{N} \longrightarrow B$ such that the following two squares commute:

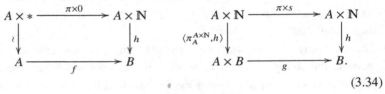

$$(3.34)$$

Letting $A = \mathbb{N}^m$ and $B = \mathbb{N}$ we can see that these squares commuting is the same as saying Equations (3.26) and (3.27) are satisfied. The takeaway of this is that **PRCompN** has a natural number object.

The category **PRCompN** has products and a natural number object. Let us consider all categories, including **PRCompN**, with products and a natural number object. Functors between such categories will be assumed to preserve the product and the natural number object. That is, $F(f \times g) = F(f) \times' F(g)$ and $F(\mathbb{N}) = \mathbb{N}'$, $F(*) = *'$, $F(0) = 0'$, and $F(s) = s'$. We form the category **CatXN** of categories with products and a natural number object whose morphisms preserve the product and natural number object. (In fact, it is a 2-category.) Román proved the following theorem:

Theorem 3.41. PRCompN *is an initial object in* **CatXN**.

This means the category of primitive recursive functions is the free category with products and recursion. See Hofstra and Scott (2021) to see this result in context with many related results.

Advanced Topic 3.42. We close our discussion of primitive recursive functions with an interesting historical vignette. Primitive recursive functions were defined by David Hilbert. He believed that this type of function was what was meant by a (total) computable function. Hilbert's student Wilhelm Ackermann showed that the class of all primitive recursive functions does not contain all total computable functions. In particular, he showed that there is a morphism in **TotCompN** called the *Ackermann function* A: Nat × Nat $\longrightarrow$ Nat that is computable but is not primitive recursive. We define A as follows:

$$A(m,n) = \begin{cases} n+1 & \text{if } m = 0 \\ A(m-1,1) & \text{if } m > 0 \text{ and } n = 0 \\ A(m-1, A(m,n-1)) & \text{if } m > 0 \text{ and } n > 0. \end{cases} \tag{3.35}$$

(There is a lot of fun in programming the Ackermann function and determining its values. Try to get your computer to find the value of $A(4,4)$.) In §6.2, we will meet another function that is in **TotCompN** but not in **PRCompN**. The fact that the Ackermann function is not primitive recursive is proved in Section 4.9 of Davis et al. (1994b), Section 10.3 of Cutland (1980), and Appendix A of Eilenberg and Elgot (1970).

3.4 Manipulating Bits

While one can think of a computation as manipulating strings or numbers, the most obvious way to think of a computation is as a process that manipulates bits. After all, that's how every digital electronic computer works.

Definition 3.43. The category **CompBool** has powers of Bool* type as objects. A typical object is (Bool*)^n. The morphisms in this category are computable functions whose input and output are powers of strings of Boolean types. These functions might be partial functions.

There is a subcategory **TotCompBool** that has the same objects but whose morphisms are total computable functions. There is an obvious inclusion **TotCompBool** $\longhookrightarrow$ **CompBool**.

Example 3.44. Here are some functions in **TotCompBool** and **CompBool**.

- **Addition.** +: Bool* × Bool* $\longrightarrow$ Bool*. This function adds the two binary sequences.
- **Exclusive or.** ⊕: Bool* × Bool* $\longrightarrow$ Bool* takes two strings and does the exclusive-or operation on each bit.
- **Sequences in** π. π-Pos: Bool* $\longrightarrow$ Bool* takes a string of zeros and ones and determines the earliest position of that sequence within the binary expansion of π. A program will have to generate the bits of π and see if the desired string appears. It might not. This function is in **CompBool** but might not be in **TotCompBool**. It is actually an open question in number theory if every possible sequence occurs in the expansion of π.

There is an inclusion Inc from **CompBool** into **CompFunc** such that
$\text{Inc}((\text{Bool}^*)^n) = (\text{Bool}^*)^n$. Just as we can encode any sequence of types
as powers of strings or powers of natural numbers, so too, we can encode any
sequence of types as powers of Booleans. From this we get the analogy of
Theorems 3.15 and 3.24:

Theorem 3.45. *The inclusion functor* $\text{Inc}\colon$ **CompBool** $\longhookrightarrow$ **CompFunc** *is
injective on objects, full, faithful, and inherently surjective. The same is also
true for the inclusion functor* **TotCompBool** $\longhookrightarrow$ **TotCompFunc**.

Turing machines compute string functions and register machines compute
natural number functions. We are left with the obvious question: What physical
devices compute Boolean functions? Answer: logical circuits. We briefly
review some basics of logical circuits. They are built from ANDs, ORs, NOTs,
NANDs, and NOR which we draw as follows,

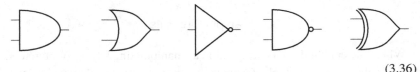

$$(3.36)$$

These gates generate all logical circuits such as the one in Figure 3.5 which has
six inputs and two outputs.

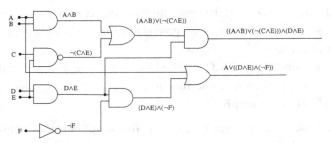

Figure 3.5 A logical circuit with 6 inputs and 2 outputs.

Logical circuits compose sequentially and in parallel. Let us look at each
one carefully.

Sequential composition is what happens when you combine a circuit with
m input wires and n output wires and another circuit with with n input wires
and p output wires. Then the outputs of the first circuit can be attached to the
input wires of the second one to form a new circuit with m input wires and p
output wires. There is a simple identity circuit which consists of a series of
small straight wires that do not have any gates. When you compose any circuit

with this identity circuit you get the same circuit with longer leads in to the input or longer leads out of the output.

Circuits also compose in parallel. That means that if there is a circuit with m input wires and n output wires and another circuit with m' input wires and n' output wires, then we can place them "side-by-side" and make a parallel circuit with $m + m'$ input wires and $n + n'$ output wires. The trivial or empty circuit with no wires is the unit for the parallel composition.

Exercise 3.46. Show that both compositions are associative.

While circuits are great, there is a major problem with such a simple model of computation. Let us say we wanted to describe a computable function $f: \text{Bool}^* \longrightarrow \text{Bool}^*$ using a logical circuit. How many input wires and output wires does one need? Obviously, this depends on the size of the inputs and outputs to the function. If the input is five bits long and there are 20 output bits, then the circuit must have five input wires and 20 output wires. On the other hand, if the input is 50 bits and the output is 501 bits long, then we will need a circuit with 50 input wires and 501 output wires. The point is that in order for a single function to be computed by logical circuits, we do not just need one circuit but a whole family or class of circuits. Each member of the family will be doing the job for a different size of input. Computing functions of the form $f: (\text{Bool}^*)^m \longrightarrow (\text{Bool}^*)^n$ is even more complicated. Here each one of the m inputs and the n outputs can be any size. That means for everyone of the the $m + n$ sizes there will be a different circuit. Let us be formal about this.

Definition 3.47. A *circuit family* with m inputs and n outputs, written $\{C\}_n^m$ is a class of circuits indexed by $m + n$ natural numbers. Each of the elements of the class will be denoted by $C_{o_1,o_2,\ldots o_n}^{i_1,i_2,\ldots i_m}$, where each of the i_j and o_j is a natural number. Such an element is drawn as follows:

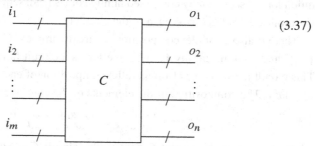

$$(3.37)$$

We furthermore demand a uniformity condition. Each of the infinite class of circuits needs to be somewhat similar and uniform in the following sense: there is a computable function that takes the $m + n$ indices and outputs the design of that circuit. (For technical reasons we further require that this function works

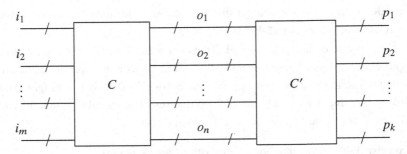

Figure 3.6 Sequential composition of circuits.

as a "log-space transducer". This means that after the input is entered, not more than logarithmic space is needed to output the circuit design. We will understand "log-space" when we finish §5.3.)

Just as there are sequential and parallel compositions of circuits, there are also sequential and parallel compositions of circuit families. Given two circuit families $\{C\}_n^m$ and $\{C'\}_k^n$ one can sequentially compose them to get $\{C' \circ C\}_k^m$. This works like the usual sequential composition of logical circuits for the indices that match up. That means for the n natural numbers $o_1, o_2, \ldots o_n$ we have circuits

$$C_{o_1,o_2,\ldots o_n}^{i_1,i_2,\ldots i_m} \quad \text{and} \quad C_{p_1,p_2,\ldots p_k}^{\prime o_1,o_2,\ldots o_n}, \tag{3.38}$$

and then they can be combined by attaching the output of the first with the input of the second to form

$$(C' \circ C)_{p_1,p_2,\ldots p_k}^{i_1,i_2,\ldots i_m}. \tag{3.39}$$

Sequentially attaching two circuits is shown in Figure 3.6. The uniformity condition for the sequentially composed family comes from sequentially composing the programs that describe each of the circuits.

There is also a parallel composition of circuit families. Given a circuit family $\{C\}_n^m$ and a circuit family $\{C'\}_{n'}^{m'}$, one forms a circuit family $\{C \otimes C'\}_{n+n'}^{m+m'}$. This circuit family comes from parallel composition of each of the elements of the class. The composition of the elements of the class is symbolized as

$$C_{o_1,o_2,\ldots o_n}^{i_1,i_2,\ldots i_m} \otimes C_{k_1,k_2,\ldots,k_{n'}}^{\prime j_1,j_2,\ldots j_{m'}} = C_{o_1,o_2,\ldots o_n,k_1,k_2,\ldots,k_{n'}}^{\prime\prime i_1,i_2,\ldots i_m,j_1,j_2,\ldots,j_{m'}}, \tag{3.40}$$

where C'' is just the circuit C next to the circuit C'. See Figure 3.7. The uniformity condition for the parallel composed families comes from the parallel composition of the programs that describe the circuits.

Exercise 3.48. Show that circuit families satisfy an interchange law. This

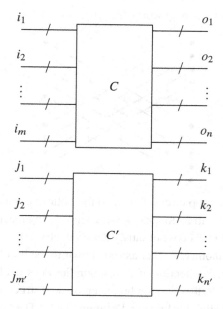

Figure 3.7 Parallel composition of circuits.

means that given four circuit families $\{C_1\}_n^m$, $\{C_2\}_k^n$, $\{C_3\}_{n'}^{m'}$, and $\{C_4\}_{k'}^{n'}$, the following rule holds:

$$(\{C_2\}_k^n \circ \{C_1\}_n^m) \otimes (\{C_4\}_{k'}^{n'} \circ \{C_3\}_{n'}^{m'}) = (\{C_2\}_k^n \otimes \{C_4\}_{k'}^{n'}) \circ (\{C_1\}_n^m \otimes \{C_3\}_{n'}^{m'}).$$

Both sides describe a circuit $\{C\}_{k+k'}^{m+m'}$.

Similar to Technical Points 3.19 and 3.27, we have the following.

Technical Point 3.49. Neither sequential nor parallel composition of circuit families is associative. The reason is that the composition of the programs that describe the circuits is not associative. There is also a problem of identity morphisms. How long will the wires be for the identity circuit? As small as they are, when we compose them with another circuit, we do not get the exact same circuit back. Rather we get a different circuit that performs the same function. By contrast, the trivial circuit family does, in fact, give a unit for the parallel composition of circuits. Because of problems with associativity and units, the collection of all circuit families does not form a category, but a bicategory.

There is one further problem with circuits that we did not find with Turing machines and register machines.

Technical Point 3.50. There is a twisting map tw that takes two circuit families $\{C\}_n^m$ and $\{C'\}_{n'}^{m'}$ and switches their order. That is, tw takes $\{C \otimes C'\}_{n+n'}^{m+m'}$ to

$\{C' \otimes C\}_{n+n'}^{m+m'}$. The twist map works like this:

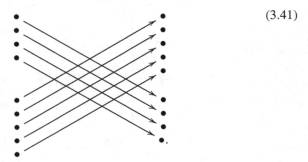

(3.41)

This crossing takes a top circuit family to the bottom circuit family and vice versa. However, in contrast to what we saw with Turing machines and register machines, this twist map does not satisfy the symmetry axiom in Definition 2.13 rather they satisfy another, weaker axiom similar to the braiding axiom of that definition. Hence the collection of circuit families does not have a symmetric monoidal structure. Rather, it has a braided monoidal structure. The reason for this is that we were slightly sloppy in Diagram (3.41). The categories of circuit families fail to be symmetric monoidal categories for a very physical reason. The physical wires cross either as

(3.42)

One sees the failure of the symmetry axiom by looking at the following

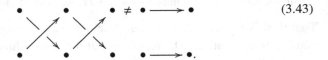

(3.43)

Now that we have all the details, let us summarize what we saw about the collection of all circuit families.

Definition 3.51. The collection of all circuit families form a braided monoidal bicategory called **Circ**. The objects are the natural numbers (which is exactly the case for **Turing** and **RegMach**.) The morphisms from m to n will be all circuit families with m input indices and n output indices. The composition will correspond to sequential composition of circuit families. The monoidal structure corresponds to parallel composition of circuit families. The iso 2-cells will be generated by associativity and unit conditions as we saw with Turing machines and register machines. The monoidal structure has a braiding.

There is a braided monoidal sub-bicategory with the same objects, and with only circuit families that do not go into an infinite loop called **TotCirc**[4]. There is an obvious inclusion **TotCirc** $\longhookrightarrow$ **Circ**.

There is a functor P: **Circ** $\longrightarrow$ **CompBool** from the braided monoidal bicategory of circuit families **Circ** to the symmetric monoidal category of computable functions on powers of Bool*, **CompBool**, that takes m to $(\text{Bool}^*)^m$ and takes a circuit family with m input indices and n output indices to the computable function from $(\text{Bool}^*)^m$ to $(\text{Bool}^*)^n$ that the circuit family describes. Iso 2-cells in **Circ** go to identity morphisms in **CompBool**. The Church–Turing thesis says that this functor is full.

The functor P describes a congruence on the category **Circ**. The circuit family $\{C\}_n^m$ is equivalent to $\{C'\}_n^m$ if they both describe the same Boolean function, i.e., $P(\{C\}) = P(\{C'\})$.

It is well known that every Boolean function h: $(\text{Bool}^*)^m \longrightarrow (\text{Bool}^*)^n$ can be computed by a circuit with only NAND gates and the fanout operation. Another way to say this is that $P^{-1}(h)$ contains a circuit family with only NAND gates and the fanout operation. Yet another way of saying this is that every circuit family in **Circ** is equivalent to a circuit with only NAND gates and fanout operations. Categorically we can say that there is a braided monoidal sub-bicategory NAND**Circ** whose circuit families are generated by NAND gates and hence includes into **Circ**. There is a functor the other way as in the following diagram:

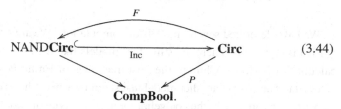

$$(3.44)$$

The two triangles commute, and $F \circ \text{Inc} = \text{Id}$. But, in general, $\text{Inc} \circ F \neq \text{Id}$.

We showed that all these different models of computation are essentially the same by showing that they all map onto **CompFunc**. They are also connected in other ways. The literature has many methods of going from one model of computation to another. For example, in Section 9.3 of Sipser (2006) there

---

[4]Circuits without any feedback produce total functions. The input goes in, and since there is no feedback, there will most definitely be output. We deal with such total functions in §5 when we discuss complexity theory. There, we always have a computable solution and we only want to talk of efficiency of computing. However, if we want to deal with the possibility of partial functions, we have to discuss feedback. While theoretical computer scientists usually do not discuss circuits that can go into an infinite loop, the computer sitting on your desk or the computer that you are reading this document on, has such a feedback mechanism and hence can go into an infinite loop. It is because of this ubiquity of circuits with feedback in the real world that we include it here.

is essentially a functor from Turing machines to circuits. In Chapters 5 and Chapter 6 of Davis et al. (1994b) there is essentially a functor from the category of register machines to the category of Turing machines. One can think of any compiler as a functor from the category of a higher-level programming language to machine language, or to circuits. In all these cases there are procedures to show how to go from one computational model to another. In order to show that these procedures are actually functors, we would have to show that they respect composition. In order to show that they are actually symmetric monoidal functors, we would have to show that they respect parallel composition. There is a lot of work! We will not do any of it, but you should be aware that one can define functors between any two models of computation such that all the small triangles in the Figure 3.8 commute. There is no reason to think that the large outer triangles should commute.

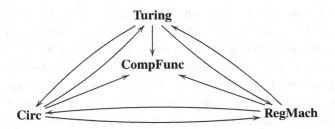

Figure 3.8 Functors between models of computation.

We have described several models of computation. We are not claiming that our list is complete. There are still other models like the λ-calculus ("lambda-calculus") of Alonzo Church, the Post machines of Emile Post, and cellular automata. There is yet another model that should warm the heart of any category theorist: one can look at the coalgebra for a certain type of endofunctor that has the information of a Turing machine. We will content ourselves by saying that it can be shown that all these models of computation map onto the same category of computable functions.

Let us pause for a moment of philosophical reflection. All these different models of computation were formulated independently from the 1930s on. Various researchers had an intuition of what a computation is and came up with their own model. As we have seen, a lot of work went in to showing that all these different models of computation all describe the *exact same category* of computable functions. Whatever is computable in this model is computable in that model as well. (In a sense this is the ultimate statement of the Church–Turing thesis.) One almost gets the impression that the category of computable functions somehow has an independent existence and that all of the various researchers independently stumbled upon this category. This is an example of the philosophical doctrine called platonism or realism, the belief that abstract objects actually exist independent of human observation or invention. The example of computable functions is a mildly convincing argument for platonism. While I know of arguments that this coincidence does not prove the philosophical conclusion of platonism, and I do not personally agree with platonism and the metaphysical gobbledygook that it entails, I think this example gives one pause for thought.

3.5 Logic and Computation

There is yet another way of describing a computation: logical formulas. Logic is the language of exact thinking and a computation is an exact process. Hence every computation can be described with a set of logical formulas. (There are many different types of logic that are used to describe computation. Here we restrict ourselves to classical propositional logic.)

We will have to build up to the definition of **Logic**. Let's start with some basics. A logical formula is a sentence built up from an ordered set of variables and logical operations $\wedge$ (conjunction), $\vee$ (disjunction), $\neg$ (negation), $\rightarrow$ (implication), and $\leftrightarrow$ (bi-implication). An example of a logical formula is $\psi = A \rightarrow (\neg(B \wedge \neg C) \vee (B \leftrightarrow D))$. We will need finite sequences of logical formulas.

Example 3.52. Here are three examples of sequences of logical formulas.

- $\Psi_1 = A \vee B \vee A \vee C, \quad (B \leftrightarrow \neg C) \vee (E \wedge (\neg(D \wedge C)))$.
- $\Psi_2 = (Z \wedge \neg Y) \leftrightarrow Z, Y, \neg Z, Y, Y$.
- $\Psi_3 = P \leftrightarrow (Q \wedge U), \quad (R \vee \neg P) \rightarrow S$.

Ψ_1 has five variables and two formulas, Ψ_2 has two variables and five formulas, and Ψ_3 has five variables and two formulas.

One can think of a sequence of logical formulas as a function. If there are m variables and n formulas, then by giving truth values into the m variables, one gets n output truth values. In effect the sequence of logical formulas with m variables and n formulas describe a function $\text{Bool}^m \longrightarrow \text{Bool}^n$. This puts logical formulas into the framework of all the other functions and processes we have seen in this text.

There are two types of compositions of sequences of logical formulas. We formulated them by looking at the composition of circuits. Let us look at both compositions carefully.

Parallel composition is easier to describe: make sure the variable names are disjoint from each other and simply concatenate the sequences. For example

- $\Psi_1 \otimes \Psi_2 = A \vee B \vee A \vee C, ((B \leftrightarrow \neg C) \vee (E \wedge (\neg(D \wedge C))), (Z \wedge \neg Y) \leftrightarrow Z, Y, \neg Z, Y, Y.$
- $\Psi_2 \otimes \Psi_1 = (Z \wedge \neg Y) \leftrightarrow Z, Y, \neg Z, Y, Y, A \vee B \vee A \vee C, (B \leftrightarrow \neg C) \vee (E \wedge (\neg(D \wedge C))).$
- $\Psi_2 \otimes \Psi_3 = (Z \wedge \neg Y) \leftrightarrow Z, Y, \neg Z, Y, Y, P \leftrightarrow (Q \wedge U), (R \vee \neg P) \rightarrow S.$
- $\Psi_3 \otimes \Psi_2 = P \leftrightarrow (Q \wedge U), (R \vee \neg P) \rightarrow S, (Z \wedge \neg Y) \leftrightarrow Z, Y, \neg Z, Y, Y.$

Notice that if Ψ has m variables and n formulas, and Ψ' has m' variables and n' formulas, then $\Psi \otimes \Psi'$ has $m + m'$ variables and $n + n'$ formulas. The unit of parallel composition is the empty sequence with no variables.

Sequential combination is slightly more complicated. We can sequentially compose Ψ, which has m variables and n formulas, with Ψ', which has n variables and k formulas, to form $\Psi' \circ \Psi$, which has has m variables and k formulas. The way this is done is that the ith formula of Ψ becomes the ith variable of Ψ'. For example, $\Psi_2 \circ \Psi_1$ is the sequence

- $[(B \leftrightarrow \neg C) \vee (E \wedge (\neg(D \wedge C))) \wedge \neg(A \vee B \vee A \vee C)] \leftrightarrow (B \leftrightarrow \neg C) \vee (E \wedge (\neg(D \wedge C))),$
- $(B \leftrightarrow \neg C) \vee (E \wedge (\neg(D \wedge C))),$
- $\neg(A \vee B \vee A \vee C),$
- $(B \leftrightarrow \neg C) \vee (E \wedge (\neg(D \wedge C))),$
- $(B \leftrightarrow \neg C) \vee (E \wedge (\neg(D \wedge C))).$

While the theory of sequences of logical formula is perfectly sound, it is not sophisticated enough to work for us. For the same reasons we needed families of logical circuits in §3.4, we need families of sequences of logical formulas. To describe a computable function $f : \mathsf{Bool}^* \longrightarrow \mathsf{Bool}^*$ using a sequence of logical formulas, we need to take into account all the possible sizes of the inputs and the outputs. Things get even more complicated when we try to describe a computable function $f : (\mathsf{Bool}^*)^m \longrightarrow (\mathsf{Bool}^*)^n$. We must take into account all the possible sizes of the m inputs and all the possible sizes of the n outputs.

Definition 3.53. A *family of sequences of logical formulas* $\{\Psi\}_n^m$ with m variable indices and n formula indices is a family or class of sequences of logical formulas indexed by $m + n$ natural numbers with a uniformity condition which says that given the natural number indices a computer can produce the sequence of logical formulas.

The rest of the discussion follows along the same lines that were done in §3.4 when we discussed $\{C\}_n^m$. It does not pay to repeat it for $\{\Psi\}_n^m$.

Similar to Technical Points 3.19, 3.27, and 3.49, we have the following.

Technical Point 3.54. The associativity of sequential and parallel composition fails for two reasons. When we change the the names of overlapping variables, there are problems. Furthermore, there are problems when composing the programs for the uniformity condition. There are similar issues with the units of the compositions.[5]

Let us summarize what we know about the collection of all families of sequences of logical formulas.

Definition 3.55. There is a symmetric monoidal bicategory **Logic** of families of sequences of logical formulas. The objects are the natural numbers. The morphisms from m to n is the set of all families of sequences of logical formulas with m variable indices and n formula indices. Sequential and parallel composition are as described above. The iso 2-cells are generated by associations and unit situations. The twisting map simply switches the order of the two sequences.

There is a functor from the braided monoidal bicategory **TotCirc** to the symmetric monoidal bicategory **Logic**,

$$L_c : \textbf{TotCirc} \longrightarrow \textbf{Logic}, \qquad (3.45)$$

that can describe a computation as a family of sequences of logical formulas. Here, L_c is the identity on objects (natural numbers) and takes the circuit to the logical formula it describes. For example, the circuit in Figure 3.5 will correspond to the sequence of logical formulas

$$((A \wedge B) \vee (\neg(C \wedge E))) \wedge (D \wedge E), \quad A \vee ((D \wedge E) \wedge (\neg F)). \qquad (3.46)$$

The way to see that L_c is a symmetric monoidal functor is to realise that the category **Logic** was formed by imitating how the category of circuits without feedback was formed.

---

[5]The fact that the collection of "physical" machines like Turing machines, register machines, and circuits are not easily described by classical categories is understandable. After all, categories were created to deal with collections of mathematical ideal structures and these "physical" machines are not totally idealistic. (Neither are they truly "physical" because Turing machines have non-physical infinite tapes, register machines can hold arbitrary large natural numbers, and circuit families have to deal with arbitrary large number of inputs.) However, the fact that the collection of families of sequences of logical formulas also fails to be easily described by classical categories came somewhat of a surprise. The problem stems from the fact that variables are used. Even if one accepts some platonism/realism point of view in the philosophy of mathematics, one must accept that in Plato's "attic" where – supposedly – all mathematical structures, including all families of sequences of logical formulas are neatly laid out, there are no variables. They simply are not part of the idealistic reality of mathematics (if such a thing exists). They are part of the presentation of mathematics. And all presentations are inherently messy.

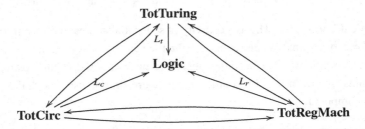

Figure 3.9 Functors between models of computation and logical formulas.

One can go on and compose L_c with a functor **TotRegMach** $\longrightarrow$ **TotCirc** which gives us L_r: **TotRegMach** $\longrightarrow$ **Logic**. We can also compose L_c with a functor **TotTuring** $\longrightarrow$ **TotCirc** which gives us L_t: **TotTuring** $\longrightarrow$ **Logic**. There are many such functors from total Turing machines and total register machines to logical formulas. Chapter 11 of Boolos et al. (2007) has them both. Section 2.6 of Garey and Johnson (1979) has a nice reduction of Turing machines to logical formulas. See also Section 6.5 of Cutland (1980) for a reduction of register machines to logical formula. All these functors fit into Figure 3.9's diagram of symmetric and braided monoidal bicategories and functors.

These functors express the same relationship between computational models and sequences of logical formulas. We shall state it for total Turing machines. If T is a total Turing machine then $L_t(T)$ is a family of sequences of logical formulas. If L_x is either L_t, L_c, or L_r and M is the appropriate model of computation, then $L_x(M)$ is a family of sequences of logical formulas. Furthermore, for every input w for M, we have that $L_x(M)[w]$ is a sequence of logical formulas that will be satisfiable if and only if model M accepts input w.

Both **Circ** and **Logic** map onto computable Boolean functions. How are all three of these categories related? There is an important congruence relation that one can put on the category **Logic**. Let us start at the beginning. We say two logical formulas ψ and ψ' are "logically equivalent" if, when we put in the same truth values into the variables, they have the same truth values. Notice that ψ and ψ' need not have the same variables, just the same number of variables. If ψ and ψ' are logically equivalent we write $\psi \sim \psi'$. The relation $\sim$ forms an equivalence relation on the set of all logical formulas.

Let us move the discussion from logical formulas to sequences of logical formulas. Two sequences of logical formulas Ψ and Ψ' are logically equivalent, written $\Psi \sim \Psi'$, when they have the same number of formulas and each formula in Ψ is logically equivalent to its corresponding formula in Ψ'. Notice that $\sim$

forms an equivalence relation on the set of all sequences of logical formulas.

Exercise 3.56. Show that $\sim$ respects sequential composition and hence is a congruence relation. That is, if $\Phi \sim \Psi$ and $\Phi' \sim \Psi'$ then $\Phi \circ \Phi' \sim \Psi \circ \Psi'$.

Exercise 3.57. Show that $\sim$ respects parallel composition. That is, if $\Phi \sim \Psi$ and $\Phi' \sim \Psi'$ then $\Phi \otimes \Phi' \sim \Psi \otimes \Psi'$.

Let us now move the discussion from sequences of logical formulas to families of sequences of logical formulas. We say $\{\Psi\}_n^m$ is logically equivalent to $\{\Psi'\}_n^m$, written $\{\Psi\}_n^m \sim \{\Psi'\}_n^m$, when each corresponding sequence is equivalent. This forms an equivalence relation and, following the last two exercises, forms a congruence on the symmetric monoidal bicategory **Logic**. We can form the quotient symmetric monoidal bicategory **Logic**/$\sim$ whose objects are natural numbers and whose morphisms are equivalence classes of families of sequences of logical formulas. There is a symmetric monoidal functor π: **Logic** $\longrightarrow$ **Logic**/$\sim$ that is the identity on objects and takes every family of sequences of logical formula to its equivalence class. By dealing with equivalent families of sequences of logical formulas, we are talking about a symmetric monoidal category as opposed to a bicategory.[6]

There is also an isomorphism of symmetric monoidal categories

$$S\colon \mathbf{CompBool} \longrightarrow \mathbf{Logic}/\sim \tag{3.47}$$

that takes every Boolean function $f\colon \mathsf{Bool}^m \longrightarrow \mathsf{Bool}^n$ to the equivalence class of sequences of logical formulas that describe this computable function.

Exercise 3.58. Show that S is an isomorphism of symmetric monoidal categories.

In summary we have the following commutative diagram

$$\begin{array}{ccc}
\mathbf{TotCirc} & \xrightarrow{\;\;L_c\;\;} & \mathbf{Logic} \\
{\scriptstyle P}\downarrow & & \downarrow{\scriptstyle \pi} \\
\mathbf{TotCompBool} & \xrightarrow[\;\;S\;\;]{\;\sim\;} & \mathbf{Logic}/\sim
\end{array} \tag{3.48}$$

where the top row has symmetric monoidal bicategories and the bottom row has symmetric monoidal categories.

---

[6]The quotient category is denoted **Logic**/$\sim$ and should be called the "weak Lindenbaum–Tarski category." We call it *weak* because families of sequences of logical formulas are set to be equivalent only if they describe the same functions *and* they have the same number of input variables and the same number of formulas. Presumably, the Lindenbaum–Tarski category should be able to describe when two families of sequences of formulas are equivalent even when the number of variables are different and the number of formulas are different. At the moment, I do not know how to make a category of these equivalence classes.

Advanced Topic 3.59. This connection between computation and logic is the beginning of fields called *descriptive complexity theory* and *finite model theory*. These fields measure how complicated certain structures are by how complicated their logical descriptions are. (See Immerman, 1998, and Libkin, 2004.)

Research Project 3.60. Here is a research project that is a little more philosophical in nature. Different models of computation satisfy different conditions of being a symmetric monoidal bicategories. As we saw, the collection of computable functions satisfy all the axioms of being a symmetric monoidal category. Why is it that some models satisfy more axioms than others? Why is it that pure mathematical objects satisfy more axioms than physical objects? See the footnote on page 53.

3.6 Numbering Machines and Computable Functions

Now that we finished introducing all the models of computation, we need to have a little chat about encoding the models. Let us go back and discuss Turing machines. The set of states of a Turing machine, its alphabet, and rules can all be encoded into a finite natural number. The intuition behind this is that the information can be encoded as a string and then the string can be encoded into a binary number (using ASCII, for example). That binary number is the number of the Turing machine. Another way to see this is to realise that every computer program can be stored as a sequence of zeros and ones, which can be thought of as a binary number. The numbers for most Turing machines will be astronomically large, but that does not concern us.

If we were to get down to the details of actually encoding a Turing machine, we would be a few ingredients. First we need a function that encodes symbols as numbers (like the ASCII code). We also need to pair two numbers as one. An example of a *pairing function* which is $\langle , \rangle \colon$ Nat $\times$ Nat $\longrightarrow$ Nat, defined by $\langle m, n \rangle = 2^m (2n + 1) - 1$. To understand this function, notice that *the fundamental theorem of arithmetic* (also called *the unique-prime-factorization theorem*) says that every positive integer can be written uniquely as a product of prime numbers. That is, for all $y > 0$ there is a unique sequence of natural numbers $x_1, x_2, \ldots, x_t$ such that $y = 2^{x_1} 3^{x_2} 5^{x_3} \cdots p_t^{x_t}$. Another way to say this is that every integer can be uniquely written as a power of 2 (say 2^{x_1}) and an odd number (say $3^{x_2} 5^{x_3} \cdots p_t^{x_t}$). Since we want the pairing function to be an isomorphism, and hence demand 0 to be in the image of the pairing function, we subtract 1. Another important function that describes a sequence of numbers as a unique number directly again uses the fundamental theorem of arithmetic.

This is the function $[?, ?, ?, \ldots, ?]\colon \text{Nat}^* \longrightarrow \text{Nat}$ defined by

$$[x_1, x_2, \ldots, x_t] = 2^{x_1} 3^{x_2} 5^{x_3} \cdots p_t^{x_t} - 1. \tag{3.49}$$

The subtraction of 1 is, again, because we want to have 0 in the image of the function. The fact that this is an isomorphism is a restatement of the fundamental theorem of arithmetic.

Putting all these ingredients together we can describe an isomorphism Enc that encodes a Turing machine as a unique natural number. It is very important that both Enc and Enc^{-1} are total computable functions. That means that given a Turing machine T, we can easily find its number $\text{Enc}(T)$, and given a number y we can easily find the yth Turing machine $\text{Enc}^{-1}(y) = T_y$. We do not require Enc to be functorial. That means, we do not insist that the encoding should respect the compositions of Turing machines. So rather than requiring that the the encoding is a functor from **Turing** to the category of natural numbers, we have it as a isomorphic function (or a functor of discrete categories)

$$\text{Enc}\colon \coprod_m \coprod_n \text{Hom}_{\textbf{Turing}}(m, n) \longrightarrow d(\mathbb{N}) \tag{3.50}$$

from the discrete set of all Turing machines to the discrete set of natural numbers. There are many possible computable encoding functions.

So far we have talked about encoding Turing machines. What about encoding computable functions? The full functor **Turing** $\longrightarrow$ **CompFunc** induces an onto functor

$$\coprod_m \coprod_n \text{Hom}_{\textbf{Turing}}(m, n) \longrightarrow \coprod_{\text{Seq}} \coprod_{\text{Seq}'} \text{Hom}_{\textbf{CompFunc}}(\text{Seq}, \text{Seq}') \tag{3.51}$$

which fits into the commutative diagram

$$\coprod_m \coprod_n \text{Hom}_{\textbf{Turing}}(m, n) \tag{3.52}$$

$$\coprod_{\text{Seq}} \coprod_{\text{Seq}'} \text{Hom}_{\textbf{CompFunc}}(\text{Seq}, \text{Seq}') \xrightarrow{\quad\sim\quad} d(\mathbb{N})$$

Notice that the bottom functor is not an isomorphim but a surjection. This functor provides a way of talking about the number of a computable function. If T is a Turing machine and $\text{Enc}(T) = y$ then the function that T performs will be called φ_y. Similarly, if φ is a computable function, then there is a Turing machine T that computes it. If $\text{Enc}(T) = x$ then we say that φ is the xth computable function. Since there are many Turing machines that describe the same computable function (there are, in fact, countably infinite Turing machines for every computable function), there are going to be many numbers for every function. This means that the enumeration $\varphi_0, \varphi_1, \varphi_2, \varphi_3, \ldots$ contains every single computable function (with infinite repetitions.)

Let us do some counting. Since every Turing machine is associated with a natural number, there are countably infinite ($\aleph_0$) Turing machines. There are also countably infinite computable functions in **CompString**. For every computable function on strings there are countably infinite Turing machines that compute that computable function (we can always make changes to a Turing machine.) In contrast, there are uncountably infinite ($2^{\aleph_0}$) functions in **Func**. While computer scientists usually focus on what can be computed, it pays to keep in mind that the vast majority of functions cannot be computed by any computing device.

What we just said about encoding Turing machines is true about all the other computational models we encountered. They are all finite machines or structures that can be specified with a unique finite number of finite strings, rules, programs, etc. This means that we can encode all of them. For every natural number n we can talk about the nth register machine M_n or the nth family of logical circuits, $\{C\}_n$, etc. We can look at all of these encodings in one diagram:

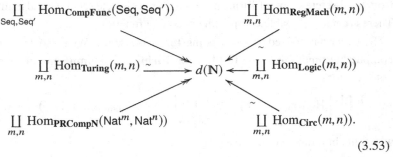

$$(3.53)$$

Notice that all the encodings of descriptions of computable functions are isomorphims. In contrast, the encodings of computable functions and subcategories such as primitive recursive functions are surjections but not isomorphims.

All these encodings are called *Gödel numbering* or *Gödelization* in honor of Kurt Gödel who was the first person to number such structures. He was interested in giving a number to every logical formula. He gave a number to a logical formula that describes numbers. He was then able to have formulas talk about formulas and numbers talk about numbers. Gödel was making logic self-referential. We number programs so that we can have programs talk about programs. This self-reference will be very helpful throughout this text and especially in §6.

These encodings arise in a very important theorem. Every computable function can be encoded as a number. Consider the computable function φ_e.

This means that Turing machine T_e describes φ_e, i.e., we insist that the process of going from e to T_e is computable. This means that there is a computable function that can take an input e and find the Turing machine T_e and then simulate T_e on any given input.

Theorem 3.61. *There exists a computable function called the universal Turing machine U : Nat $\times$ String $\longrightarrow$ String such that for all natural numbers e and string inputs w, we have $U(e, w) = \varphi_e(w)$.*

We don't formally prove the theorem here. For a proof, see the texts on computability theory at the end of this section and the next.

The function U can compute all functions. (This is analogous to an operating system that is a computer program that can execute any computer program.)

With the encoding of programs and functions we can also demonstrate the close relationship between programs and inputs. Take a computable function φ_e that demands $m + n$ inputs, say $x_1, x_2, \ldots, x_m, y_1, y_2, \ldots, y_n$. For convenience we can think of these inputs as numbers. It is possible to "hard wire" the x_i of the inputs into a function; that is, make the x_i of the inputs part of the program. Consider the computable function that demands only n inputs that works as follows:

```
1. Move contents of input tape 1 to input tape m+1 and put
   x₁ on input tape 1.
2. Move contents of input tape 2 to input tape m+2 and put
   x₂ on input tape 2.
3. Move contents of input tape 3 to input tape m+3 and put
   x₃ on input tape 3.
   ⋮
n. Move contents of input tape n to input tape m+n and put
   xₙ on input tape n.
n+1. Execute Turing machine e.
```

This Turing machine has a number. Notice that, given $x_1, x_2, \ldots, x_m$ and e, the encoding of this Turing machine is totally computable. We call the number of this Turing machine $s_n^m(e, x_1, x_2, \ldots, x_m)$. The above constitutes the proof of the following important theorem.

Theorem 3.62 (The SMN theorem). *For every m and n, there is a total computable function $s_n^m : Nat^{m+1} \longrightarrow Nat$ such that for every computable function $\varphi_e : Nat^{m+n} \longrightarrow Nat$, we have*

$$\varphi_{s_n^m(e, x_1, x_2, \ldots, x_m)}(y_1, y_2, \ldots, y_n) = \varphi_e(x_1, x_2, \ldots, x_m, y_1, y_2, \ldots, y_n). \quad (3.54)$$

This theorem is related to the fact that the category of computable functions, **CompFunc**, is Cartesian closed (i.e, functions of the form $A \longrightarrow C^B$ are equivalent to functions of the form $A \times B \longrightarrow C$.) This is also related to the programming concept called "currying."

Research Project 3.63. The usual encodings do not necessarily respect the composition, i.e., $\text{Enc}(T' \circ T) \neq \text{Enc}(T') + \text{Enc}(T)$. Perhaps there does exist some encodings where the composition is respected. Perhaps one can prove that no such encoding exists.

Further Reading

Every book in theoretical computer science has their favorite model of computation.

- Many books use Turing machines for historical reasons. Sipser (2006), Lewis and Papadimitriou (1997), and Boolos et al. (2007) all use Turing machines. There is more about the development of the Turing machine idea in Andrew Hodges' excellent biography of Alan Turing (Hodges, 1983).
- Register machines can be found in Cutland (1980); Davis et al. (1994b); Rogers (1987).
- Circuits are usually found in discussions of complexity theory. See for example Sipser (2006), Cormen et al. (2009), and Arora and Barak (2009). As we mentioned, if we permit circuits with feedback that can go into an infinite loop, then we can also discuss computability theory.
- The relationship between computation and logic is widely covered. See in particular Chapters 9–11 of Boolos et al. (2007), Chapter 4–6 of Papadimitriou (1994), and Chapters 12 and 13 of Davis et al. (1994a). None of them deal with multiple inputs and outputs so none of them directly mention our construction of families of sequences of logical formulas.

4 Computability Theory

This section deals with the question of which functions are computable, and more interestingly, which functions are not computable. We will be stating our theorems in terms of the top spoke of *The Big Picture*. The categories **TotCompString** and **CompString** will not play a role here and so we will concentrate on the following part:

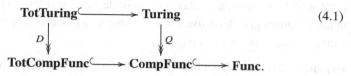

$$(4.1)$$

A large part of our discussion of computability theory is determining if a given morphism in **Func** is in **CompFunc** or in **TotCompFunc**. Another way of looking at this is to consider the following functors

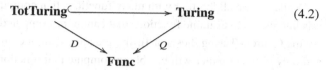

$$(4.2)$$

and ask if a particular morphism in **Func** is in the image of Q, or in the image of D, or neither. (Obviously, if it is in the image of D, then it is also in the image of Q.)

4.1 Turing's Halting Problem

There are various names for morphisms in **Func**.

Definition 4.1. A function in **TotCompFunc** is termed *totally computable* or *total Turing computable* or *totally solvable*. Analogously, a function in **CompFunc** is called *computable* or *Turing computable* or *solvable*. A function that is in **CompFunc** but not in **TotCompFunc** is also called a *partially computable* function.

Morphisms whose codomain is Bool have special names. The map $f : Seq \longrightarrow$ Bool is called a *decision problem*. An instance of the problem is input into the function and the output is either true or false. If the decision problem is in **TotCompFunc** (i.e., there is a total Turing machine that always outputs true or false), it is called *recursive*, or *Turing-decidable*, or simply *decidable*. In computability theory there are decision problems for which computers cannot give such definitive answers. Such decision problems have some inputs that get a true answer and have some inputs that go on and on and never halt. In contrast to a "true–false" answer, we might say they give a "true–maybe" answer. (Of course, the "maybe" might be eternal silence.) These are very important decision problems in **CompFunc** called *recursively enumerable*, or simply *r.e.*[7]. That means that, for a given input to the function, a Turing machine can

---

[7]Within the literature, these decision problems go by many different names. They are also called: computably enumerable, c.e., Turing-recognizable, and semi-decidable.

recognize when the answer is true, but might not be able to tell when the answer is false. (There are even weaker notions of decision problems where a Turing machine cannot give definitive answers either way.) Decision problems will be a major focus in the coming pages.

Example 4.2. All the functions in §§3.2, 3.3, and 3.4 for which Turing machines, register machines, or circuits were described are examples of computable functions. The functions that we showed were primitive recursive or recursive are also examples of computable functions.

While we are all familiar with many functions that computers can compute, it is interesting to examine functions that cannot be computed by any computer. By the Church–Turing thesis, a Turing machine cannot compute a function if and only if no computer will be able to compute that function. Let us describe one such.

We are interested in determining when a Turing machine halts and when a Turing machine goes into an infinite loop. In many cases, whether or not the Turing machine halts depends on its input. Let us simplify the problem by considering Turing machines that only take a single natural number as input. There is a morphism in **Func** called $\mathrm{Halt} \colon \mathrm{Nat} \times \mathrm{Nat} \longrightarrow \mathrm{Bool}$ which is defined as follows

$$\mathrm{Halt}(x, y) = \begin{cases} 1 : & \text{if Turing machine } y \text{ on input } x \text{ halts.} \\ 0 : & \text{if Turing machine } y \text{ on input } x \text{ does not halt.} \end{cases}$$

Let us spend a moment "unpacking" what this function does. As we saw in §3.6, every Turing machine can be described by a unique number. We might write the Turing machine whose number is y as T_y. Furthermore, we are interested in Turing machines that accept a single number in their input tape. The Halt function is determining if Turing machine T_y halts when it gets the the single number x as input, or if Turing machine T_y goes into an infinite loop when it gets the single number x as input.

The *Halting problem* or the *Halting decision problem* asks if one is able to write a Turing machine to compute the Halt function. The map Halt is a total function, but is it in **TotCompFunc**, **CompFunc** or only in **Func**?

Theorem 4.3 (Turing's undecidability of the Halting problem). *The* Halt *function is not in* **TotCompFunc**. *That is,* Halt *is not recursive (or Turing-decidable).*

Proof. First some intuition. The proof is an example of a self-referential paradox, a famous example of which being the "Liar paradox" which is the sentence "This sentence is false." That sentence is true if and only if it is false, i.e., it is a contradiction. Here, with the Halting problem, the same type of self-referential statement is made as follows:

> If there was a way to solve the Halting problem, then one can construct a program that performs the following task: "When you ask this program if it will halt, then it will give the wrong answer."

Since computers do not give wrong answers, this program does not exist and hence the Halting problem is unsolvable.

In detail, the proof is a proof by contradiction. Assume (wrongly) that Halt is in **TotCompFunc**. We will compose Halt with two morphisms

- the diagonal morphism, Δ: Nat $\longrightarrow$ Nat $\times$ Nat, in **TotCompFunc** defined by $\Delta(n) = (n, n)$, and
- the "partial Not" morphism, ParNOT: Bool $\longrightarrow$ Bool, in **CompFunc** (not in **TotCompFunc**) defined as

$$\text{ParNOT}(x) = \begin{cases} 1: & \text{if } x = 0 \\ \uparrow: & \text{if } x = 1. \end{cases} \tag{4.3}$$

where $\uparrow$ means go into an infinite loop.

After composing as follows

$$\text{Nat} \xrightarrow{\ \Delta\ } \text{Nat} \times \text{Nat} \xrightarrow{\ \text{Halt}\ } \text{Bool} \xrightarrow{\ \text{ParNOT}\ } \text{Bool}$$

with the arc labeled Halt' from Nat to Bool.

we obtain Halt' which is in **CompFunc** since all of the morphisms it is composed of are in **CompFunc** (by assumption).

Halt' is defined as

$$\text{Halt}'(x) = \begin{cases} 1: & \text{if Turing machine } x \text{ on input } x \text{ does not halt.} \\ \uparrow: & \text{if Turing machine } x \text{ on input } x \text{ does halt.} \end{cases} \tag{4.4}$$

Since Halt' is in **CompFunc** there is some Turing machine, say T_{y_0}, that computes Halt'. Let us ask Halt' about itself by plugging y_0 into Halt'. Then Halt'(y_0) halts and outputs a 1 if and only if Turing machine y_0 on input y_0 does not halt but goes into an infinite loop, i.e.,

$$\text{Halt}'(y_0) = 1 \iff \text{Halt}'(y_0) = \uparrow. \tag{4.5}$$

This is a contradiction. We asked T_{y_0} about itself and we got the wrong answer. The only thing we assumed was that Halt was in **TotCompFunc**. We conclude that Halt is not in the subcategory **TotCompFunc** of **Func**. $\square$

In contrast to the total function Halt, there is a partial halting function Parhalt: Nat × Nat ⟶ Bool defined as

$$\text{Parhalt}(x, y) = \begin{cases} 1: & \text{if Turing machine } y \text{ on input } x \text{ halts.} \\ \uparrow: & \text{if Turing machine } y \text{ on input } x \text{ does not halt.} \end{cases}$$

Theorem 4.4. Parhalt *is recursively enumerable (or Turing-recognizable).*

Proof. The idea is that there exists a Turing machine that can give a true–maybe answer to the halting question. This Turing machine – let us call it The Simulator – will accept the number y and will use the computable encoding function of §3.6 to figure out the rules of Turing machine T_y. With these rules in hand, The Simulator will perform the operations of T_y or, in other words, simulate T_y. (This is similar to an operating system that performs the operations of a program, or simulates a program.) The Simulator will simulate T_y when it gets number x. If, while simulating, Turing machine y halts on input x, The Simulator will output a 1. In contrast, as long as the simulation of Turing machine y on input x does not halt, The Simulator will not give any response and will carry on. ☐

Since we are discussing partial ways to solve the Halting problem, there is another function that will be important later. The timed halting function TimeHalt: Nat × Nat × Nat ⟶ Bool is defined as

$$\text{TimeHalt}(x, y, t) = \begin{cases} 1: & \begin{array}{l} \text{if Turing machine } y \text{ on input } x \text{ halts} \\ \qquad \text{within } t \text{ steps.} \end{array} \\ 0: & \begin{array}{l} \text{if Turing machine } y \text{ on input } x \text{ does not halt} \\ \qquad \text{within } t \text{ steps.} \end{array} \end{cases}$$

Theorem 4.5. TimeHalt *is in* **TotCompFunc***, i.e., it is recursive (or Turing-decidable).*

Proof. The idea is that there exists a Turing machine that can give a true–false answer to the timed halting question. This Turing machine – let us call it The Simulator too – will accept the number y and will simulate the T_y Turing machine on input x. The Simulator will also be keeping a counter of how many steps it has simulated. If, while simulating, Turing machine y halts on input x within t steps, then The Simulator will output a 1. In contrast, once The Simulator passes step t, it will output a 0. ☐

Notice that the following logical statement is true.

$$\text{Halt}(x, y) = 1 \text{ if and only if } \exists t \, \text{TimeHalt}(x, y, t). \tag{4.6}$$

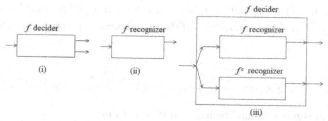

Figure 4.10 (i) a decider, (ii) a recognizer, and (iii) a decider built out of two recognizers

Let us find a morphism that is not even in **CompFunc**. First a definition and theorem. The function NOT: Bool $\longrightarrow$ Bool is defined as NOT(0) = 1 and NOT(1) = 0, and is obviously in **TotCompFunc**.

Theorem 4.6. *Let f: Seq $\longrightarrow$ Bool be in* **CompFunc** *and let*

$$f^c = NOT \circ f: Seq \longrightarrow Bool \longrightarrow Bool. \qquad (4.7)$$

Then f is in **TotCompFunc** *if and only if f and f^c are in* **CompFunc**, *i.e., are both recursively enumerable.*

Proof. If f is in **TotCompFunc**, then f is definitely in **CompFunc**. Since NOT: Bool $\longrightarrow$ Bool is in **TotCompFunc** it follows that $f^c = $ NOT $\circ$ f is in **TotCompFunc** and hence in **CompFunc**.

The other direction of the proof is slightly more complicated. One can gain intuition by looking at the three parts of Figure 4.10. In (i) we see a decider function that gives a true–false answer. The input enters on the left and either true or false is marked on the right. In (ii) we have a recognizer function. The input enters on the left and the function recognizes true answers and outputs true on the right. However, the function might not be able to tell when the input does not have a property. In that case there is no answer. Part (iii) of the figure shows how one can build a decider by using two recognizers. The input is entered on the left and it goes into two recognizers. Both recognizers are executed in parallel. Since one of them is true, one of them will answer true.

In detail, assume that f and f^c are in **CompFunc**. The function that will be used to parallel process those two functions at one time will be the function Parallel: Bool $\times$ Bool $\longrightarrow$ Bool defined as

$$\text{Parallel}(x, y) = \begin{cases} 1: & \text{if } x = 1 \\ 0: & \text{if } y = 1. \end{cases} \qquad (4.8)$$

For the way we are using this function, it will never be the case that $x = y = 1$ or $x = y = 0$ so we do not mind what value Parallel gives for those inputs

(let it be 0). The Parallel morphism is a total morphism, and hence is in
TotCompFunc. The composition of the morphisms in **CompFunc** is given as

$$\text{Seq} \xrightarrow{\Delta} \text{Seq} \times \text{Seq} \xrightarrow{f \times f} \text{Bool} \times \text{Bool} \xrightarrow{\text{id} \times \text{NOT}} \text{Bool} \times \text{Bool} \xrightarrow{\text{Parallel}} \text{Bool} .$$

with the top arc labeled $f \times f^c$.

This morphism is total and hence in **TotCompFunc**. $\square$

Now let us use this theorem to consider the complement of the partial halting
function

$$\text{Parhalt}^c = \text{NOT} \circ \text{Parhalt}: \text{Nat} \times \text{Nat} \longrightarrow \text{Bool} \longrightarrow \text{Bool} \qquad (4.9)$$

defined by

$$\text{Parhalt}^c(x, y) = \begin{cases} 1: & \text{if Turing machine } y \text{ on input } x \text{ does not halt.} \\ \uparrow: & \text{if Turing machine } y \text{ on input } x \text{ halts.} \end{cases}$$

Theorem 4.7. Parhalt^c *is in* **Func** *but not in* **CompFunc**, *i.e., it is not even
recursively enumerable.*

Proof. If Parhalt and Parhalt^c were both in **CompFunc** then from Theorem 4.6
we would be able to combine them to have *Halt* in **TotCompFunc**. We know
that this is not true from Theorem 4.3. This is intuitively true: while we can
give a positive response when a Turing machine halts, how can we ever give a
positive response that a Turing machine *will never* halt? $\square$

4.2 Other Unsolvable Problems

While the Halting problem is undecidable, it is just the beginning of the story.
There are many other decision problems that are as undecidable and hard or
harder than the Halting problem. They too are undecidable. But first we need a
way of comparing decision problems.

Definition 4.8. A *reduction* is a way of discussing the relation between two
decision problems. Let $f: \text{Seq} \longrightarrow \text{Bool}$ and $g: \text{Seq}' \longrightarrow \text{Bool}$ be two functions
in **Func**. We say that f is *reducible* to g or f *reduces* to g if there exists an
$h: \text{Seq} \longrightarrow \text{Seq}'$ in **TotCompFunc** such that

$$\text{Seq} \xrightarrow{h} \text{Seq}' \qquad (4.10)$$

with f and g mapping down to Bool.

commutes. We write this as $f \leq g$. If $f \leq g$ and $g \leq f$ then we write $f \equiv g$ and say they are both part of the same *computability class*.

The way to think about such a reduction is that h changes an f input into a g input. Letting x be the input to f, the commuting triangle requirement means

$$f(x) \text{ is true if and only if } g(h(x)) \text{ is true.} \qquad (4.11)$$

Why is the notion of reduction important? Note that if there is a way to solve g then there is definitely a way to solve f: simply use h to change the input of f into an input of g and then solve it. The contrapositive of this statement is central: If there is no way to solve f then there is no way to solve g. Another way to say this is that the g decision problem is as hard or harder than the f decision problem.

A categorical way to view reducibility is to consider the following two functors

$$\textbf{TotCompFunc} \xhookrightarrow{\text{Inc}} \textbf{Func} \xleftarrow{\text{Const}_{\text{Bool}}} \textbf{1} \qquad (4.12)$$

where the left functor is the inclusion and the right functor picks out the type Bool. We then form the comma category $(\text{Inc}, \text{Const}_{\text{Bool}})$. The objects of this category are decision problems in **Func** and the morphisms are total computable functions that make Diagram (4.10) commute. It is important to note that this is not a slice category because we only want special types of morphisms between objects. In other words, we want only totally computable reductions between decision problems.

Let us use this notion of reducibility to prove that some morphisms are like Halt and are not in **TotCompFunc**.

Example 4.9. The *Nonempty program problem* asks if a given (number of a) Turing machine will have a nonempty domain. This means that there is some input into the Turing machine which the Turing machine will accept. There is a morphism in **Func** called Nonempty: Nat $\longrightarrow$ Bool which is defined as follows

$$\text{Nonempty}(y) = \begin{cases} 1: & \text{if Turing machine } y \text{ has a nonempty domain} \\ 0: & \text{if Turing machine } y \text{ has an empty domain.} \end{cases}$$

We show that the Halting problem reduces to the Nonempty program problem as in

$$\text{Nat} \times \text{Nat} \xrightarrow{h} \text{Nat} \qquad (4.13)$$

with Halt and Nonempty mapping to Bool.

The total computable function h is defined as follows for Turing machine y and for input x: $h(x,y) = y'$ where y' is the number of the Turing machine that performs the following task:

<u>Turing machine y':</u>

on input w

1. If $w \neq x$ reject and stop.
2. If $w = x$ execute Turing machine y on input x. If Turing machine y accepts x, accept and stop.

Notice that Turing machine y' depends on x and y. Function h is easily seen to be totally computable. That means that a Turing machine can easily describe y' if it is given x and y. Now consider Turing machine y'. It is basically a Turing machine that has at most one element in its domain. Only number x can possibly be in its domain. Furthermore, whether or not the domain of y' is empty is dependent on whether or not x is accepted by Turing machine y. Formally,

$$\text{Nonempty}(y') = 1 \iff \text{Nonempty}(h(x,y)) = 1 \text{ by definition of } h$$
$$\iff \text{the domain of Turing machine } y' \text{ is not empty}$$
$$\iff x \text{ is in the domain of Turing machine } y'$$
$$\iff \text{Turing machine } y \text{ accepts } x$$
$$\iff \text{Halt}(x,y) = 1.$$

But we already know that it is impossible to solve the Halting problem. So it must be impossible to solve the Nonempty problem.

Example 4.10. The opposite of the Nonempty program problem is the *Empty program problem*. This decides if the domain of (the number of) a given Turing machine is empty. The Empty program problem is undecidable because if it was decidable, then we would be able to compose it with the NOT: Bool $\longrightarrow$ Bool to get a decider for the Nonempty program problem

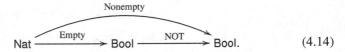

$$\text{Nat} \xrightarrow{\text{Empty}} \text{Bool} \xrightarrow{\text{NOT}} \text{Bool}. \tag{4.14}$$

Since we know that Nonempty is not computable, we know that Empty is not computable.

Example 4.11. The *Equivalent program problem* asks if two given (numbers of) Turing machines describe the same function. That is, if Turing machine T and T' always give the same output for the same input. There is a morphism in

Func called Equiv : Nat × Nat $\longrightarrow$ Bool which is defined as follows

$$\text{Equiv}(y, y') = \begin{cases} 1 : & \text{if } T_y \text{ describes the same function as } T_{y'} \\ 0 : & \text{if } T_y \text{ does not describe the same function as } T_{y'}. \end{cases}$$

In order to show that Equiv is not in **TotCompFunc** we show that we can reduce Empty to Equiv as follows:

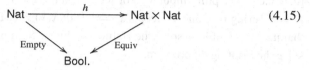

$$(4.15)$$

Let y_0 be the number of a silly Turing machine that simply goes into an infinite loop for any input. Nothing is ever accepted. This machine clearly has an empty domain, i.e., $\text{Empty}(y_0) = 1$. Now we shall use this Turing machine to describe a reduction from Empty to Equiv, namely h : Nat $\longrightarrow$ Nat × Nat which is defined for Turing machine y as $h(y) = (y, y_0)$. Notice that $\text{Equiv}(y, y_0) = 1 \Longleftrightarrow$ Turing machine y performs the same function as the silly Turing machine $y_0 \Longleftrightarrow \text{Empty}(y) = 1$. But since we know that Empty is not computable, we know that Equiv is not computable.

Example 4.12. The *Printing* 42 *problem* asks if a given (number of a) Turing machine has some input for which 42 is an output. There is a morphism in **Func** called Print : Nat $\longrightarrow$ Bool which is defined as follows

$$\text{Print}(y) = \begin{cases} 1 : & \text{if there exists an input to } T_y \text{ that outputs 42.} \\ 0 : & \text{if there does not exist an input to } T_y \text{ that outputs 42.} \end{cases}$$

(Obviously the number 42 is not important to the problem. It is simply the answer to the ultimate question of life, the universe, and everything.)

We show that the Halting problem reduces to the Printing 42 problem as in

$$\text{Nat} \times \text{Nat} \xrightarrow{\;\;h\;\;} \text{Nat} \qquad\qquad (4.16)$$

$$\text{Halt} \searrow \qquad \swarrow \text{Print}$$

$$\text{Bool.}$$

The total computable function h is defined for Turing machine y and for input x as $h(x, y) = y'$ where y' is the number of the Turing machine that performs the following task:

Turing machine y':

on input w

1. If $w \neq x$ reject and stop.

2. If $w = x$ execute Turing machine y on input x. If Turing
   machine y accepts x, print "42", accept and stop.

Notice that Turing machine y' depends on x and y. Also notice that function h is easily seen to be totally computable. That means that a program can easily compute y' if it is given x and y. Now consider Turing machine y'. It accepts at most one word which is x. Whether or not y' prints 42 depends on Turing machine y accepting input x. Formally, $\text{Print}(y') = 1 \iff \text{Print}(h(x,y)) = 1 \iff$ Turing machine y accepts $x \iff \text{Halt}(x,y) = 1$. But we already know that it is impossible to solve the Halting problem. So it must be impossible to solve the Printing 42 problem.

There are many other decision problems that can be shown to be undecidable. In fact we will show that a computer cannot deal with the vast majority of the properties of Turing machines. What type of properties are we talking about? First we are dealing with nontrivial properties. By this we mean that there exist Turing machines that have the property and Turing machines that do not have the property. It is very easy to decide trivial properties (just always answer 0 or always answer 1.) We are also interested in semantic properties. By this we mean we are interested in properties of the function that the Turing machine produces. In other words, if Turing machine y produces the same function as Turing machine y', then both y and y' have a semantic property or both do not have a semantic property. In contrast to a semantic property, a syntactical property of Turing machines is very easy to decide. For example, it is easy to decide if a Turing machine has 100 rules or more. Similarly it is easy to decide if a Turing machine uses less than 37 states. A Turing machine can be written to answer such questions.

Theorem 4.13 (Rice's theorem). *Any nontrivial, semantic property of Turing machines is undecidable.*

Proof. Let P be a nontrivial, semantic property. There will be a morphism in **Func** that decides property P, i.e., a function Pdecider: Nat $\longrightarrow$ Bool where $\text{Pdecider}(y) = 1 \iff$ Turing machine y has property P. We show that the Halting problem is reducible to the problem of deciding the P property with the map h_P

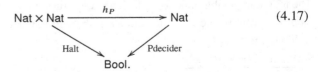

$$\text{(4.17)}$$

Let us say that y_0 is the number of the silly Turing machine that always rejects every input. Either Turing machine y_0 has property P or does not have property

P. Assume that it does not (the proof can easily be modified if Turing machine y_0 does have property P). Since P is nontrivial there exists a Turing machine y_1 that does have property P. So we have Pdecider(y_0) = 0 and Pdecider(y_1) = 1. We define $h_P(x, y) = y'$ where y' is the number of the following Turing machine:

```
Turing machine y':
on input w
```

1. Simulate Turing machine y on input x.

    (a) If it halts and rejects, then reject and stop.
    (b) If it halts and accepts, then go to step 2.

2. Simulate Turing machine y₁ on w.

Notice that if Turing machine y rejects x then Turing machine y' rejects any input and will be equivalent to Turing machine y_0. Also, if Turing machine y on input x goes into an infinite loop, then w will not be accepted, just as Turing machine y_0. In contrast, if Turing machine y on input x accepts, then Turing machine y' will act just as Turing machine y_1. Formally,

$$\text{Pdecider}(y') = 1 \leftrightarrow \text{Pdecider}(h_P(x, y)) = 1$$
$$\leftrightarrow \text{Turing machine } h_P(x, y) = y' \text{ acts like Turing machine } y_1$$
$$\leftrightarrow \text{Halt}(x, y) = 1.$$

But we know that the Halting problem cannot be solved. We conclude that Pdecider is not computable. □

Example 4.14. Here is just a small sample of the nontrivial semantic properties that are not decidable by Rice's theorem:

- Determine if a Turing machine has a finite domain.
- Determine if a Turing machine has an infinite domain.
- Determine if a Turing machine accepts a particular input.
- Determine if a Turing machine accepts all inputs.

Gödel's incompleteness theorem is one of the most important theorems of 20th century mathematics. The theorem can be seen as a simple consequence of the undecidability of the Halting problem. It would be criminal to be so close to it and not state and prove it.

First some preliminaries. We say a logical system is *complete* if every statement that is true has a proof within the system. In contrast, a logical system is *incomplete* if there exists a statement that is true for which there is no proof within the system. A system is *inconsistent* if there is some proposition P such that both P and $\neg P$ are proved. A system is *consistent* if it is not inconsistent.

Theorem 4.15 (Gödel's Incompleteness Theorem). *For any consistent logical system which is powerful enough to deal with basic arithmetic, there are statements that are true but unprovable. That is, the logical system is incomplete.*

Proof. This is more a sketch of a proof. Since we are dealing with an exact logical system where the axioms are clear and the method of proving theorems are exact, it is possible for a computer to decide when a string is a formal proof of the logical system. This follows from the fact that if the logical system is able to perform basic arithmetic, its statements can be encoded as numbers. Hence, it is possible (though extremely inefficient) to produce all strings in lexicographical order, and to have a computer determine if the string is a formal proof of a statement or the negation of a statement. This amounts to saying that there is a computable function SuperEval: String $\longrightarrow$ Bool that evaluates a logical formula ϕ and tells us if it or its negation is provably true. It is defined as

$$
\text{SuperEval}(\phi) = \begin{cases} 1 : & \text{if there exists a proof that } \phi \text{ is true} \\ 0 : & \text{if there exists a proof that } \neg\phi \text{ is true.} \end{cases} \tag{4.18}
$$

The main question is whether the computable function SuperEval is total or not. In a complete logical system, SuperEval is total. In contrast, in an incomplete logical system, there exists a statement ϕ such that neither ϕ nor $\neg\phi$ have proofs, and hence SuperEval is not total.

We will not go through all the details, however much of the proof has been set up already when we discussed the functor L_t from the category of total Turing machines to the category of families of sequences of logical formulas on page 54. Remember that, for a total Turing machine T and an input w there is a sequence of logical formulas $L_t(T)[w]$ that describes a potential computation of Turing machine T with input w. Total Turing machine T on input w is accepted if and only if the conjunction of the logical formulas in the sequence $L_t(T)[w]$ is satisfiable. This functor can be extended to $\widehat{L}_t$ from the symmetric monoidal category of all Turing machines to the category **Logic'** of all families of sequences of predicate logic. In **Logic'** we have formulas with quantifiers $\forall$ and $\exists$. This functor will describe computations that might not halt with predicate logic. The functor $\widehat{L}_t$ has the following property: Turing machine T on input w halts and accepts if and only if $\widehat{L}_t(T)[w]$ is not only true but provable. We will use this fact to reduce the Halting problem to a logical problem.

We will use a reduction from Halt to show that SuperEval is not total and the system is incomplete. There is a total computable function h: Nat $\times$ Nat $\longrightarrow$

String that makes the following triangle commute

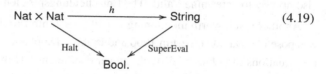

$$h(x, y) = \widehat{L_t}(T_y)[x]. \qquad (4.20)$$

This means that $h(x, y)$ will produce the sequence of logical formulas that is a string of symbols which logically describes the potential computation of T_y on input x. Now h is a computable function because the functor $\widehat{L_t}$ is a computable function. Moreover $h(x, y)$ is true if and only if $\widehat{L_t}(T_y)[x]$ is provable if and only if Turing machine y halts on input x. One can think of $h(x, y)$ as the predicate formula $\exists z \text{TimeHalt}(x, y, z)$. If SuperEval was total, then we would be able to solve the Halting problem. Since we know the Halting problem is not solvable, it must be that SuperEval is not total and hence the logical system is not complete. □

There is another logical result that is worth mentioning. Alan Turing's original paper is titled "On computable numbers, with an application to the Entscheidungsproblem," (Turing, 1937). The Entscheidungsproblem or "decision problem" is a question that David Hilbert asked. He wanted to know if there exists some algorithm that can decide if any given logical formula is valid (true no matter what variables are used.) Both Alan Turing and his thesis adviser, Alonzo Church, independently showed that there is no such algorithm.

Theorem 4.16. *The Entscheidungsproblem is unsolvable.*

Turing showed that if there was a computable function Valid : String $\longrightarrow$ Bool which accepts a logical formulas and is defined as

$$\text{Valid}(\phi) = \begin{cases} 1 : & \text{if } \phi \text{ is valid} \\ 0 : & \text{if } \phi \text{ is not valid.} \end{cases} \qquad (4.21)$$

then one can reduce Halt to Valid as follows.

Nat × Nat ────────h────────→ String $\qquad (4.22)$

Halt ↘ ↙ Valid

Bool.

While we will not go through the details of the proof, the h in this case is related to the h in Equation (4.20). Since there is no way to determine Halt, there is also no way to determine Valid. The Entscheidungsproblem is unsolvable.

Another result worth mentioning is Hilbert's Tenth Problem. This problem was posed by David Hilbert in 1900 and is about Diophantine equations. These are equations of polynomials with integer coefficients. Hilbert's 10th problem asks if there is a computable way to determine if any given Diophantine equation has integer solutions. Over the next 70 years many researchers like Martin Davis, Hilary Putnam, and Julia Robinson made progress on this problem. Finally in 1970, a 22-year old named Yuri Matiyasevich proved there is no computable way to determine Hilbert's 10th problem. The proof basically shows that if there was a computable function $\text{Dioph}: \text{String} \longrightarrow \text{Bool}$ that accepts a Diophantine equation and is defined as

$$
\text{Dioph}(p = p') = \begin{cases} 1: & \text{if } p = p' \text{ has integer solutions} \\ 0: & \text{if } p = p' \text{ does not have integer solutions} \end{cases}
$$

then there is a reduction from Halt to Dioph:

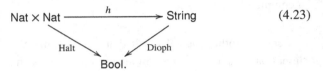

$$(4.23)$$

Since Halt is not computable, then Dioph is not computable.

Advanced Topic 4.17. Researchers have studied special types of decision problems on natural numbers. They look at functions of the form $f: \text{Nat} \longrightarrow \text{Bool}$ which correspond to subsets of natural numbers. By examining the computablity of the characteristic function, they have defined many types of subsets of numbers with fanciful names such as "recursive," "productive," "creative," "simple," "hypersimple," "r.e.," "immune," "hyperimmune," etc. There are uncountably infinite subsets of natural numbers, and they come in many forms. For more see Chapters 2 and 5 of Soare (1987), Chapters 7–9 of Rogers (1987), and Chapter 7 of Cutland (1980).

Research Project 4.18. It would be nice to categorically formulate the characteristic functions to describe the types of subsets mentioned in Advanced Topic 4.17.

4.3 Classifying Unsolvable Problems

What is beyond **CompFunc**? We have shown that there are morphisms that are in **Func** and not in **CompFunc**. While we have given a few examples of such functions, it is important to remember what we said on page 58 that the vast majority of morphisms in **Func** are not in **CompFunc**. While we tend to think mostly of what is in **CompFunc**, there are vastly more morphisms in **Func** that are not in **CompFunc**.

Is there a way to characterize and classify the morphisms in **Func**? This was a question that, again, goes back to Alan Turing. He showed us how to answer it by describing more powerful types of computation.

Let $f:$ Seq $\longrightarrow$ Seq$'$ be any function in **Func** (one should think of f as *not* being in **CompFunc**.) An f-*oracle Turing machine* is a Turing machine that can "magically" use f in its computation. In detail, such a Turing machine has an extra "query tape" and an extra "query state." While the Turing machine is executing, it can place some information, x, on the query tape. Once the question is in place, the Turing machine can go to the special query state. At that time click the oracle will magically erase x from the query tape and put $f(x)$ on the tape. The computation then continues on its merry way with this new piece of information. The intuition is that the computation that uses f is more powerful than a regular computation.

For any morphism f in **Func** we formulate the category of f-oracle Turing machines which we denote **Turing**$[f]$. (The notation should remind a mathematician of taking a ring and adding in an extra variable to get a larger ring.) The objects of **Turing**$[f]$ are the natural numbers. The morphisms are f-oracle Turing machines. A regular Turing machine can be thought of as an oracle Turing machine where the query tape and the query state are never used. This means that there is an inclusion functor **Turing** $\longrightarrow$ **Turing**$[f]$. We can also discuss which functions can be computed by a Turing machine that has access to the f-oracle. This gives us the category **CompFunc**$[f]$. If a function does not use the oracle, then it is in **CompFunc**. Hence there is an inclusion **CompFunc** $\longrightarrow$ **CompFunc**$[f]$. Every morphism in **CompFunc**$[f]$ is still a morphism in **Func**. We can summarize all these categories with this commutative diagram

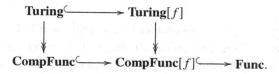

Exercise 4.19. Show that f is in **CompFunc**$[f]$.

Exercise 4.20. If f is in **CompFunc**, then show **CompFunc** = **CompFunc**$[f]$.

Exercise 4.21. Show that if f is in **CompFunc**$[g]$, and g is in **CompFunc**$[h]$, then f is in **CompFunc**$[h]$.

Rather than just taking any arbitrary non-computable f for an oracle, let us take Halt: Nat×Nat $\longrightarrow$ Bool. This will result in the category **CompFunc**[Halt] which consists of all the functions that are computable if a computer has access to the Halt function. This is a larger subcategory of **Func** than **CompFunc** but not all of **Func**. We can ask whether or not a Turing machine with a Halt-oracle will halt. These Turing machines can also be enumerated and we can make a new halt function $\widehat{\text{Halt}}$: Nat × Nat $\longrightarrow$ Bool defined as

$$\widehat{\text{Halt}}(x,y) = \begin{cases} 1 : & \text{if Halt-oracle Turing machine } y \text{ on } x \text{ halts.} \\ 0 : & \text{if Halt-oracle Turing machine } y \text{ on } x \text{ doesn't halt.} \end{cases}$$

It is not hard to show that the $\widehat{\text{Halt}}$ function is not computable even with the Halt-oracle. That is, $\widehat{\text{Halt}}$ is not in **TotCompFunc**[Halt]. We can use $\widehat{\text{Halt}}$ as a new oracle and construct **CompFunc**[$\widehat{\text{Halt}}$]. This process of going from one category of functions to a larger category of functions is called the *jump operation*. We can continue this process again and again. We have the following infinite sequence of categories:

$$\textbf{CompFunc} \hookrightarrow \textbf{CompFunc}[\text{Halt}] \hookrightarrow \textbf{CompFunc}[\widehat{\text{Halt}}]$$
$$\hookrightarrow \textbf{CompFunc}[\widehat{\widehat{\text{Halt}}}] \hookrightarrow \cdots \hookrightarrow \textbf{Func}.$$

This gives us a whole lattice of categories where **CompFunc** is the bottom and **Func** is the top. This is a classification of the uncomputable functions in **Func**. We know a lot about what we cannot compute.

Advanced Topic 4.22. In 1956, an American named Richard Friedberg and a Russian named Albert Muchnik independently proved the *Friedberg–Muchnik theorem*. This states that there are (at least) two classes of morphisms, **A** and **B**, such that

$$\tag{4.24}$$

$$\textbf{CompFunc} \xrightarrow{\quad \textbf{A} \quad} \textbf{CompFunc}[\widehat{\text{Halt}}] \xrightarrow{\quad} \textbf{Func}.$$

Furthermore, you cannot obtain all the morphisms of **B** if you use any morphism of **A** as an oracle, and you cannot obtain all the morphisms of **A** if you use any morphism of **B** as an oracle. This means that **A** and **B** are "computationally independent" of each other. This result opened the door to the study of the all the computability classes between **CompFunc** and **Func**. See Section 10.2 of Rogers (1987) and Section 7.2 of Soare (1987).

Research Project 4.23. Think of the lattice of subcategories of **Func**. It would be nice to formulate the jump operation as a endofunctor on this category. Its seems that the jump operator is some type of closure operator. Exactly what are its properties?

Further Reading

There are many excellent books on computability theory, e.g., Sipser (2006); Cutland (1980); Davis et al. (1994b).

More can be found about particular topics here:

- Unsolvable problems and Rice's Theorem: Chapter 8 of Hopcroft and Ullman (1979), Chapter 8 of Davis et al. (1994b), and Chapter 5 of Lewis and Papadimitriou (1997).
- Gödel's Incompleteness Theorem: Section 6.2 of Sipser (2006), page 23 of Arora and Barak (2009), and Chapter 6 of Papadimitriou (1994).
- Hibert's 10th Problem: Davis and Hersh (1973), Chapter 6 of Cutland (1980).
- Oracle computation and the whole hierarchy of unsolvable problems: in Soare (1987) and Rogers (1987).

5 Complexity Theory

One of the most important branches of theoretical computer science is complexity theory. Within this field we learn how to measure the amount of resources needed to solve problems. There are connections to complexity theory in almost every branch of mathematics and applied science.

While computability theory deals with what can and cannot be computed, complexity theory deals with what can and cannot be computed *efficiently*. Here we do not ask which functions are computable, but rather, which computable functions can be computed with a reasonable amount of resources. We also classify different types of computable functions by their different levels of efficiency or complexity.

Historically, complexity theory has only dealt with total computable functions. Most of our discussion will only deal with the following functor from *The Big Picture* D: **TotTuring** $\longrightarrow$ **TotCompFunc**. Remember that D takes the object m to the object String^m, and a total Turing machine with m input tapes and n output tapes goes to the total computable function, String^m $\longrightarrow$ Strings^n, it computes.

5.1 Measuring Complexity

When we say a function uses an efficient amount of resources, we usually mean the number of steps for a Turing machine to compute the function is efficient. This corresponds to the amount of time it takes for the computation to complete. The more steps needed to complete the computation, the more time will be required. Researchers have also been interested in how much computational space is needed to make a computation. Let us focus on the amount of time a computation needs first, and we will explore space requirements later in §5.3.

Usually the amount of time needed depends on two things: (i) the size of the input; one expects that a large input would demand a lot of time and a small input rather less. And (ii) the state of the input. For example, if one is interested in sorting data, then usually, data that is already almost sorted, does not require a lot of computing time to get the data fully ordered. In contrast, if the data is totally disordered, more time is needed. We will be interested in the worst-case scenario, that is, the worst possible state of the data.

In order to formalize the notion that the number of operations and time needed is dependent on the size of the input, we associate to every total Turing machine a function from the natural numbers, $\mathbb{N}$, to the set of non-negative real numbers $\mathbb{R}^*$. The function $f: \mathbb{N} \longrightarrow \mathbb{R}^*$ will describe how many operations are needed for a computation of the Turing machine. The $\mathbb{N}$ corresponds to the size of the input, and the $\mathbb{R}^*$ corresponds to the number of operations needed. (While only whole numbers are used to describe how many operations are required, since some functions will use operations like logarithms which give real number outputs, we employ the codomain $\mathbb{R}^*$.) If n is the size of the input, then $f(n)$ is the number of operations needed to complete the computation in the worst-case scenario. That is, $f(n)$ is the maximum amount of resources needed for inputs of size n.

All this is for one Turing machine that implements a function. But there are many Turing machines that implement the same function. We are going to need to consider the best-case scenario, i.e., the most efficient Turing machine that implements a certain function. For every total computable function we will associate a function $\mathbb{N} \longrightarrow \mathbb{R}^*$ which is calculated by looking at the best possible Turing machine that solves that problem, and looking at the worst possible data that can be input to that machine.

The set of all functions of the form $\mathbb{N} \longrightarrow \mathbb{R}^*$ form a set $\mathrm{Hom}_{\mathbf{Set}}(\mathbb{N}, \mathbb{R}^*)$ which is an ordered monoid. The monoid operation, $+$, is inherited from $\mathbb{R}^*$. The unit is the function that always outputs zero. The order is also inherited from $\mathbb{R}^*$. Essentially $f \leq g$ if and only if $f(n) \leq g(n)$ for all $n \in \mathbb{N}$.

When complexity theorists compare two functions in $\mathrm{Hom}_{\mathbf{Set}}(\mathbb{N}, \mathbb{R}^*)$, they want to regard two functions as being the same if they only differ by a constant multiple. They also want to ignore what happens when the input sizes are small. This is accomplished by working with a quotient ordered monoid

$\mathrm{Hom}_{\mathbf{Set}}(\mathbb{N}, \mathbb{R}^*)/\sim$ defined as follows: $f: \mathbb{N} \longrightarrow \mathbb{R}^*$ is considered the same as $g: \mathbb{N} \longrightarrow \mathbb{R}^*$, i.e., $f \sim g$ if and only if

$$0 < \lim_{n \to \infty} \frac{f(n)}{g(n)} < \infty. \tag{5.1}$$

It is not hard to see that this relation is an equivalence relation. In fact, it is a congruence for the monoid structure.

Some notation:

- Rather than using the usual equivalence class notation, i.e., $f \in [g]$, complexity theorists use the notation $f \in \Theta(g)$ or $f = \Theta(g)$. If $f = \Theta(g)$, then we say that f grows at the same rate as g. We can write $\Theta(g)$ as the set of functions that grow the same rate as g, i.e.,

$$\Theta(g) = \left\{ f : 0 < \lim_{n \to \infty} \frac{f(n)}{g(n)} < \infty \right\}. \tag{5.2}$$

- If $f \le g$ in the quotient ordered monoid, we write $f = O(g)$ and say that f grows at the same rate or slower than g. Another way of saying this is

$$O(g) = \left\{ f : 0 \le \lim_{n \to \infty} \frac{f(n)}{g(n)} < \infty \right\}. \tag{5.3}$$

- If $f = O(g)$ and $f \ne \Theta(g)$ we write $f = o(g)$ and say that f grows slower than g. Another way to say this is

$$o(g) = \left\{ f : \lim_{n \to \infty} \frac{f(n)}{g(n)} = 0 \right\}. \tag{5.4}$$

- $O(1)$ consists of constant functions which means that a constant number of operations are needed regardless of the size of the input.

Example 5.1. Some ideas and names using this notation:

- Any polynomial is in the same equivalence class as its leading exponent. For example, $16n^5 - 31n^4 + 21.7n^2 - 23.9n - 451 = \Theta(n^5)$.
- Leading exponents make a difference. That is, for any two real numbers $0 < r_1 < r_2$, we have that $n^{r_1} = o(n^{r_2})$.
- Polynomials grow faster than logarithms. For any real number $0 < r$, we have that $\log_b(n) = o(n^r)$ for any b.
- Polynomials grow slower than exponentiation. For all real numbers $0 < r$, we have that $n^r = o(b^n)$ for any $b > 1$.

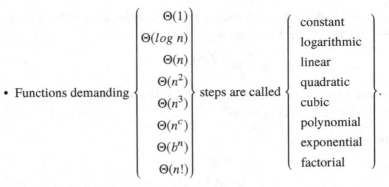

- Functions demanding $\left\{\begin{array}{c}\Theta(1)\\\Theta(\log n)\\\Theta(n)\\\Theta(n^2)\\\Theta(n^3)\\\Theta(n^c)\\\Theta(b^n)\\\Theta(n!)\end{array}\right\}$ steps are called $\left\{\begin{array}{l}\text{constant}\\\text{logarithmic}\\\text{linear}\\\text{quadratic}\\\text{cubic}\\\text{polynomial}\\\text{exponential}\\\text{factorial}\end{array}\right\}$.

Exercise 5.2. Prove the following facts about functions. (i) $O(f) - o(f) = \Theta(f)$. (ii) If $g = O(f)$ and $h = O(g)$, then $h = O(f)$. (iii) If $g = O(f)$ and $f = O(g)$, then $f = \Theta(g)$.

Before we venture on the task of measuring the resources of total computable functions, keep in mind that measuring resources is not functorial. As we will see in Technical Point 5.5, the measure of the sequential composition of two functions is not related to the measure of the functions independently. There is a similar failure for the parallel composition of two functions. Hence, many times in our discussion, we will be dealing with sets (discrete categories) and functions as opposed to (monoidal) categories and functors.

The details of how we arrive at a function that measures complexity is not simple. Feel free to skip to the punchline in Diagram (5.15) on page 82.

Step 1. Worst case data.

We need to look at all Turing machines with their appropriate input. This is constructed from the pullback

$$\begin{array}{ccc}\textbf{InpTM} & \longrightarrow & \coprod_m \coprod_n \mathrm{Hom}_{\textbf{TotTuring}}(m,n) \qquad (5.5)\\ \downarrow & & \downarrow \\ \coprod_{m=0}^{\infty}(\Sigma^*)^m & \longrightarrow & d(\mathbb{N}),\end{array}$$

which we now explain. The lower right entry is the discrete category of natural numbers. The upper right corner is the set (as opposed to the category) of all total Turing machines. The right vertical functor takes every Turing machine to the number of input string it demands. The lower left is the set of all possible input strings. The m-tuples of strings for all m is the set $\coprod_{m=0}^{\infty}(\Sigma^*)^m$. The bottom functor takes an m-tuple of strings and simply outputs m. The pullback of these two functors is the set of pairs of total Turing machines and the inputs for those Turing machines. We call this discrete category **InpTM**. The objects are pairs $\langle T, w_1, w_2, \ldots, w_m \rangle$, where T is a total Turing machine that demands m inputs, and the w_i are the inputs.

We must be able to measure the size of the input to a Turing machine. Not only do we need to measure the number of strings, but we also need to know the sum total of the lengths of all the strings. There is a length functor $|-| : \Sigma^* \longrightarrow \mathbb{N}$ that takes a string and gives the length of the string. This functor can be extended to an m-tuple of strings $|-| : (\Sigma^*)^m \longrightarrow \mathbb{N}$ in the obvious way: if $w_1, w_2, w_3, \ldots, w_m$ is an m-tuple of strings, then

$$|w_1, w_2, w_3, \ldots, w_m| = |w_1| + |w_2| + |w_3| + \cdots + |w_m|. \tag{5.6}$$

This can be further extended to m-tuples for any m. This gives us the functor $|-| : \coprod_m (\Sigma^*)^m \longrightarrow \mathbb{N}$.

Given a Turing machine and its input, there are several resources we can measure. The functor Time measures the number of computational steps that a Turing machine with an input will demand to complete the computation. The functor Space measures the number of cells needed in the work tapes to complete the computation. There are still other resources studied, for example,

$$\textbf{InpTM} \;\; \substack{\xrightarrow{\;\;\;\text{Time}\;\;\;} \\ \xrightarrow{\;\;\;\text{Space}\;\;\;} \\ \vdots \\ \xrightarrow{\;\;\;\text{others}\;\;\;}} \;\; \mathbb{R}^* \tag{5.7}$$

Every cell that is used in a computation demands one computational time step to write in that cell. However a Turing machine might write different things in a cell during the course of a computation. This means that for any particular Turing machine and an input, more time will be used than space. That is, for all $\langle T, w_1, w_2, \ldots, w_m \rangle$,

$$\text{Space}(\langle T, w_1, w_2, \ldots, w_m \rangle) \leq \text{Time}(\langle T, w_1, w_2, \ldots, w_m \rangle). \tag{5.8}$$

Categorically we say that there is a natural transformation Space $\Longrightarrow$ Time.

Let us examine important subsets of **InpTM**. For Turing machine T that demands m inputs there is a subset $\langle T, (\Sigma^*)^m \rangle$ of **InpTM** that consists of all the possible inputs to that Turing machine. Each of the measures of resources restrict to this set. From such a set, and for any measure of resource, say Time, there is a function to $\mathbb{R}^*$ and a length function to $\mathbb{N}$, i.e.,

$$\mathbb{N} \xleftarrow{\;\;\;|-|\;\;\;} \langle T, (\Sigma^*)^m \rangle \xrightarrow{\;\;\;\text{Time}\;\;\;} \mathbb{R}^*. \tag{5.9}$$

These two functions are used to define a function

$$\max_T (\text{Time}) : \mathbb{N} \longrightarrow \mathbb{R}^*. \tag{5.10}$$

This is the function that gives the time needed to compute all input. In words, $\max_T(\text{Time})$ is the function that expresses the amount of time resources used by T for the worst input w. In other words, if w is an input, look at the function g such that $g(|w|)$ is the amount of time needed. Take the maximum such g. Formally, this is

$$\max_T(\text{Time}) = \max_{\substack{g:\, \mathbb{N} \longrightarrow \mathbb{R}^* \\ w \in (\Sigma^*)^m \\ g(|w|) = \text{Time}(T,w)}} g : \mathbb{N} \longrightarrow \mathbb{R}^* \qquad (5.11)$$

There is a similar function for space:

$$\max_T(\text{Space}) = \max_{\substack{g:\, \mathbb{N} \longrightarrow \mathbb{R}^* \\ w \in (\Sigma^*)^m \\ g(|w|) = \text{Space}(T,w)}} g : \mathbb{N} \longrightarrow \mathbb{R}^* \qquad (5.12)$$

Technical Point 5.3. For those who know and love the language of Kan extensions, these max functions are simply left Kan extensions

$$(5.13)$$

The fact that $\text{Lan}_{(-)}$ is a functor and there is a natural transformation Space $\Longrightarrow$ Time means that there is an induced natural transformation $\max_T(\text{Space}) \Longrightarrow \max_T(\text{Time})$. This means that for a particular Turing machine T, its space requirements will be less than its time requirements, for any sized inputs.

Step 2: Best Case Turing machine.

We are still not done. We have found the function that describes the amount of time needed for a particular Turing machine T. It remains to find the function that describes resources needed by searching through all the Turing machines that implement some computable function. Let $f : \text{Seq} \longrightarrow \text{Seq}'$ be in **TotCompFunc**. Consider the preimage set $D^{-1}(f)$ of all Turing machines that implement f. We define the function that measures complexity of computable functions as follows:

$$\mu_{D,\text{Time}}(f) = \min_{T \in D^{-1}(f)} \max_T(\text{Time}). \qquad (5.14)$$

This gives us the desired function:

$$\text{TotCompFunc} \xrightarrow{\mu_{D,\text{Time}}} \text{Hom}_{\text{Set}}(\mathbb{N}, \mathbb{R}^*). \qquad (5.15)$$

Notice the functor D is used in the definition and hence is in the notation.

Technical Point 5.4. For those who know and love the language of Kan extensions, these min functions are simply right Kan extensions. We described above a functor

$$\max_{(-)}(\text{Time})\colon \textbf{TotTuring} \longrightarrow \text{Hom}_{\textbf{Set}}(\mathbb{N}, \mathbb{R}^*). \qquad (5.16)$$

The Kan extension is then

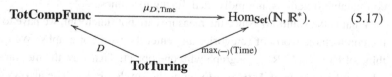

$$\textbf{TotCompFunc} \xrightarrow{\ \mu_{D,\text{Time}}\ } \text{Hom}_{\textbf{Set}}(\mathbb{N}, \mathbb{R}^*). \qquad (5.17)$$

If we use the resource Space, we would get the functor that measures space:

$$\textbf{TotCompFunc} \xrightarrow{\ \mu_{D,\text{Space}}\ } \text{Hom}_{\textbf{Set}}(\mathbb{N}, \mathbb{R}^*). \qquad (5.18)$$

The natural transformation $\max_T(\text{Space}) \Longrightarrow \max_T(\text{Time})$ induces a natural transformation

$$\textbf{TotCompFunc} \quad \Downarrow \quad \text{Hom}_{\textbf{Set}}(\mathbb{N}, \mathbb{R}^*). \qquad (5.19)$$

with arrows $\mu_{D,\text{Space}}$ (top) and $\mu_{D,\text{Time}}$ (bottom).

This says that for any function, the amount of space used is less than the amount of time used.

Technical Point 5.5. We need a little discussion about $\text{Hom}_{\textbf{Set}}(\mathbb{N}, \mathbb{R}^*)$. This set of functions has an amazing amount of structure. Algebraically, the set inherits the structure from $\mathbb{R}^*$ and hence is a monoid, group, rig (ring without negation), and it is almost a field. It is also a partial order. There are many ways to represent all this structure categorically. However, there are some issues. We are talking about measures of some aspect of totally computable functions; that is, functions of the form $\textbf{TotCompFunc} \longrightarrow \text{Hom}_{\textbf{Set}}(\mathbb{N}, \mathbb{R}^*)$. The problem is that these measuring functions are not functorial. If f and g are total computable functions, then in general we have

$$\mu_{D,\text{Time}}(g \circ f) \neq \mu_{D,\text{Time}}(g) + \mu_{D,\text{Time}}(f). \qquad (5.20)$$

A simple example where one can see this inequality is when f is a very complicated function that demands a lot of time and g is a simple function that takes every input to a single output. In this case $\mu_{D,\text{Time}}(f)$ is large, $\mu_{D,\text{Time}}(g)$ is very small, and $\mu_{D,\text{Time}}(g \circ f)$ is also very small.

Notice that – in contrast to computable functions – the time measures respect composition of Turing machines. That is, the time it takes to perform T followed by T' is equal to the time it takes to perform T plus the time it takes to perform T'. In symbols,

$$\max_{T' \circ T}(\text{Time}) = \max_{T'}(\text{Time}) + \max_{T}(\text{Time}). \qquad (5.21)$$

However, the goal of complexity theory is to measure the complexity of computable functions, not individual Turing machines.

What is the solution? Since the measures are not functorial, we have to leave the comfortable world of categories and enter the world of graphs. We want to think of $\text{Hom}_{\mathbf{Set}}(\mathbb{N}, \mathbb{R}^*)$ as a graph with one object. That way, the measures can take all the objects of **TotCompFunc** to the single object. The arrows of the graph will be the elements of $\text{Hom}_{\mathbf{Set}}(\mathbb{N}, \mathbb{R}^*)$. This needs to be a graph and not a category, because the measures do not respect the compositions of functions.

The purists among us might feign horror at having a commutative diagram where one of the categories is really a graph. Do not worry. The only construction we perform with this construction is to take pullbacks. In this sense, we are taking subsets (or subgraphs) of $\text{Hom}_{\mathbf{Set}}(\mathbb{N}, \mathbb{R}^*)$ not subcategories. And such pullbacks work just as well.

One might remedy the situation by taking the free category over the one-object graph $\text{Hom}_{\mathbf{Set}}(\mathbb{N}, \mathbb{R}^*)$. There will only be one object but the morphisms will be sequences of functions $\mathbb{N} \longrightarrow \mathbb{R}^*$. The measure will then take the composition $g \circ f$ to the sequence of measures. That is,

$$\mu_{D,\text{Time}}(g \circ f) = (\mu_{D,\text{Time}}(g), \mu_{D,\text{Time}}(f)). \qquad (5.22)$$

If the reader sleeps better at night with this remedy, then take it.

The point we are making is that measuring resources of computable functions is not functorial. There is no way to get around this dilemma.

Example 5.6. Let us examine the measures on some common total computable functions.

- **Addition.** $+: \text{Nat} \times \text{Nat} \longrightarrow \text{Nat}$. A Turing machine can add two n digit numbers on two different input tapes by adding each digit (in reverse), placing the carry digit on the work tape, a putting the result on the output tape. This does not take more than linear time. Only one digit (actually bit) is needed for space resources.

- **Multiplication.** $*: \text{Nat} \times \text{Nat} \longrightarrow \text{Nat}$. This can be done by multiplying every digit of one number by a digit of the other number. If each number is n digits long, this will demand $O(n^2)$ multiplications. The usual multiplication demands a space for carrying digits and storage space for $O(n^2)$ numbers that have to be added. There are, however, faster algorithms.

- **Searching**. Search: Nat* × Nat ⟶ Nat. Given a list of numbers and another number, the Search function returns the position of the number within the list. If the list containing n numbers is placed on one tape and the number searched for on the second tape, then the computer has to search through all the n numbers till it finds the position. The worst-case scenario is when the desired number is in the last position or does not occur in the list of numbers. In that case, the Turing machine must perform $O(n)$ comparisons. This algorithm does not use any space in the work tapes.
- **Sorting**. Sort: Nat* ⟶ Nat*. The sort function takes a string of numbers and outputs the same numbers in sorted order. There are many different sorting algorithms (See Chapters 6–8 of Cormen et al., 2009.) Many algorithms work in $O(n^2)$ and $O(n \log n)$. (For special types of input, there are algorithms that are efficient in $O(n)$ time.) There are certain algorithms that do not demand any extra space (except for a place to temporarily store a single number while swapping numbers.) Such sorts are called "in place."
- **Traveling salesman problem**.TSP: String ⟶ String. Given a weighted, complete graph as a string of characters (whose vertices are to be thought of as cities and weighted edges are to be thought of as the distance or cost of going from one city to another city), find the route that goes through every vertex exactly once with the shortest distance or the cheapest cost. The only known way to do this in all cases is to try every possible route. Since for n vertices there are $n!$ possible routes, this algorithm demands factorial time. The algorithm must store the best route found so far. So it needs to store n numbers.
- **Satisfiability problem**. SAT: String ⟶ Bool. Given a Boolean formula as a string, determine if there is some way to give values to the variables to make the complete formula true. The function returns true or false dependent on if the Boolean formula can be satisfied or not. The only way to do this in all cases is to make a truth table and see if any of the final answers are true. For a formula with n variables, there are 2^n possible rows of the truth table. Hence this algorithm demands exponential time. The algorithm needs to work out each row individually and can reuse that space for the next row. So the algorithm does not demand more space than the size of the input.

At this time, we would like to introduce another, more advanced notion of a Turing machine. Our entire discussion till now has been about *deterministic Turing machines*. These are machines that, at every time click, do exactly one operation. There are souped-up Turing machines called *nondeterministic Turing machines* that at every time click might perform one of a set of possible operations. In detail, a nondeterministic Turing machine is in state q_{32} and sees various symbols on its tapes, will have the option of doing one of a set of operations on the tapes. In analogy with Equation (3.12), we can write this as follows:

$$\delta(q_{32}, x_1, x_2, \ldots, x_t) = \{(q_{51}, y_1, y_2, \ldots, y_t, L, R, R, \ldots, L)$$
$$(q_{13}, y_1', y_2', \ldots, y_t', R, R, \ldots, L),$$
$$(q_{51}, y_1'', y_2'', \ldots, y_t'', L, R, \ldots R)\}. \tag{5.23}$$

In this example the Turing machine has three different options of operations to perform. In analogy to Diagram (3.13), the general form of the transition function for a nondeterministic Turing machine is

$$\delta : Q \times \Sigma^t \longrightarrow \mathcal{P}(Q \times \Sigma^t \times \{L, R\}^t). \tag{5.24}$$

This means that for every state q and t-tuple of inputs, there is a *whole subset of possible* states, symbols to write, and directions to go.

A computation for a nondeterministic Turing machine begins when an input is placed on the input tape and is in the start state. At every time click the Turing machine can choose one of the possible options. We say that a computation occurs when there is a sequence of choices that leads to an accepting state. The first sequence of choices that comes to the accepting state gives us a computation. The output for a nondeterministic Turing machine is the contents of the tape when the machine finds an accepting state.

There is a symmetric monoidal bicategory of nondeterministic Turing machines, **NTotTuring**. The objects are the natural numbers (as it was for **TotTuring**,) and the set of morphisms from m to n is the set of nondeterministic Turing machines with m input tapes and n output tapes. Composition of nondeterministic Turing machines happen as follows: when an accepting state is found for the first Turing machine, then the contents of its tape is the output and is considered the input of the second Turing machine tape.

Analogous to the functor from the symmetric monoidal bicategory to the symmetric monoidal category

$$D : \textbf{TotTuring} \longrightarrow \textbf{TotCompFunc} \tag{5.25}$$

there is a similar functor

$$N : \textbf{NTotTuring} \longrightarrow \textbf{TotCompFunc} \tag{5.26}$$

that takes every nondeterministic Turing machine to the computable function it computes.

The relationship between deterministic Turing machines and nondeterministic Turing machines is important. Every deterministic Turing machine can be thought of as a special type of nondeterministic Turing machine where the set of options in the transition function is a singleton set. There is an obvious inclusion functor from **TotTuring** to **NTotTuring**. The other way is more interesting. There exists a functor $F:$ **NTotTuring** $\longrightarrow$ **TotTuring** that takes every nondeterministic Turing machine to a determinstic Turing machine that performs the same computable function. The deterministic Turing machine works by systematically trying every possible computational path of the nondeterministic Turing machine until an accepting state is found. (This is admittedly not very efficient.) The fact that this functor is functorial, i.e., that

$$F(T' \circ T) = F(T') \circ F(T) \qquad (5.27)$$

can be done by making sure that the deterministic program $F(T')$ continues the systematic search of all paths in $T' \circ T$ when $F(T)$'s search is complete. With F in hand, we summarize with the following diagram:

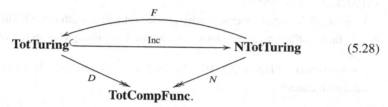

$$(5.28)$$

It should be noted that while $F \circ Inc = Id_{\text{TotTuring}}$, it is not necessarily true that $Inc \circ F = Id_{\text{NTotTuring}}$.

We discuss measuring resources by replacing the functor D with the functor N in the definition of Diagram (5.15) to get

$$\textbf{TotCompFunc} \xrightarrow{\mu_{N,\text{Time}}} \text{Hom}_{\textbf{Set}}(\mathbb{N}, \mathbb{R}^*) . \qquad (5.29)$$

The space resources measured in a deterministic Turing machine computation is the number of cells used on the work tapes. For a nondeterministic Turing machine, we measure the number of cells used in the first accepting computation. As we did with time complexity in Diagram (5.15), we can formulate the functor

$$\textbf{TotCompFunc} \xrightarrow{\mu_{N,\text{Space}}} \text{Hom}_{\textbf{Set}}(\mathbb{N}, \mathbb{R}^*) . \qquad (5.30)$$

Example 5.7. A nondeterministic Turing machine can solve the Satisfiability problem by nondeterministicly looking at every row of the truth table and evaluating the Boolean formula. This will only demand $O(n)$ time and $O(n)$ space.

With our complexity measures in our toolbox, we can go on to classify the computable functions. We look at various subsets of $\text{Hom}_{\textbf{Set}}(\mathbb{N}, \mathbb{R}^*)$ closed under addition of functions. For every such subset S (thought of as a one-object graph), we can take the following pullback and get those computable functions whose complexity is within S.

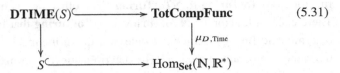

$$\text{(5.31)}$$

The objects of **DTIME**(S) are the same as **TotCompFunc**, i.e., all sequences of types, and the morphisms are those computable functions that can be computed in time that are described by some function in S. The category **DTIME**(S) is the *complexity class* of all computable functions that can be computed in S amount of time. Another complexity class is **DSPACE**(S), the collection of all computable functions that can be computed using S amount of space. Using the $\mu_{N,\text{Time}}$ and $\mu_{N,\text{Space}}$ measures, we form categories **NTIME**(S) and **NSPACE**(S), respectively.

How are these different complexity classes related to each other? There are at least three different "dimensions" to how the complexity classes are related.

- *Various subsets of* $\text{Hom}_{\textbf{Set}}(\mathbb{N}, \mathbb{R}^*)$. Given $T \subseteq S \subseteq \text{Hom}_{\textbf{Set}}(\mathbb{N}, \mathbb{R}^*)$ there is an induced inclusion

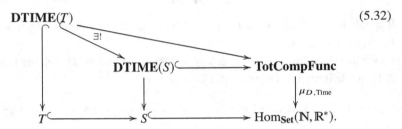

$$\text{(5.32)}$$

Similar induced inclusions exist for **DSPACE**( ), **NTIME**( ), and **NSPACE**( ).

- *Time vs. space.* Since every cell used on a Turing tape demands a time click, we have shown that for every $f: \text{Seq} \longrightarrow \text{Seq}'$ in **TotCompFunc** we have

$$\mu_{D,\text{Space}}(f) \le \mu_{D,\text{Time}}(f) \quad \text{and} \quad \mu_{N,\text{Space}}(f) \le \mu_{N,\text{Time}}(f). \quad \text{(5.33)}$$

This means that for every set S, the category of computable functions that can be done in S time can definitely be done in S space. This gives us the following inclusions:

$$\textbf{DTIME}(S) \lhook\joinrel\longrightarrow \textbf{DSPACE}(S) \quad \text{and} \quad \textbf{NTIME}(S) \lhook\joinrel\longrightarrow \textbf{NSPACE}(S).$$

- *Determinism vs. nondeterminism.* Since a deterministic Turing machine is a special type of nondeterministic Turing machine, every computable function that can be performed by a deterministic Turing machine in a certain time can also be performed by a nondeterministic Turing machine in that time. There is an induced inclusion functor from **DTIME**(S) into the category of **NTIME**(S) as in

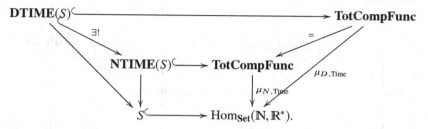

We can see all three of these dimensions in one diagram. Given $T \subseteq S \subseteq \text{Hom}_{\text{Set}}(\mathbb{N}, \mathbb{R}^*)$ we have

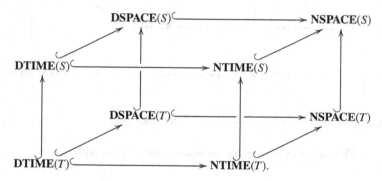

Some common subsets of $\text{Hom}_{\text{Set}}(\mathbb{N}, \mathbb{R}^*)$ that are closed under addition are Poly consisting of all functions that are polynomial or less, Const consisting of all constant functions, Exp consisting of all functions that are exponential or less, Log consisting of all functions that are logarithmic or less. These subsets are included in each other as

$$\text{Const} \hookrightarrow \text{Log} \hookrightarrow \text{Poly} \hookrightarrow \text{Exp} \hookrightarrow \text{Hom}_{\text{Set}}(\mathbb{N}, \mathbb{R}^*). \tag{5.34}$$

All these subsets induce the inclusions of complexity classes shown in Figure 5.11.

Advanced Topic 5.8. We only dealt with deterministic and nondeterministic Turing machines. There are many other types of Turing machines such as probabilistic Turing machines, quantum Turing machines, alternating Turing machines, reversible Turing machines, etc. Each with its own set of rules and with its own complexity classes. The relationship between all these complexity classes is a major topic within complexity theory. See e.g., Arora and Barak (2009); Sipser (2006); Papadimitriou (1994).

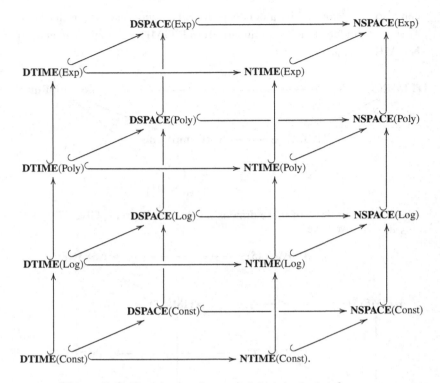

Figure 5.11 Complexity classes and their inclusion functors.

Research Project 5.9. It would be nice to see if and when these different complexity classes also have a symmetric monoidal structure.

5.2 Decision Problems

Decision problems play a major role in complexity theory. For complexity theory a decision problem is a total computable function whose codomain is Bool. We will look at the resources needed to compute these decision problems.

We saw in computability theory that reductions are a way of comparing decision problems. In complexity theory, we are interested in special types of reductions from one decision problem to another.

Definition 5.10. The reductions used in basic complexity theory are called *polynomial reduction*. Let f: Seq $\longrightarrow$ Bool and g: Seq$'$ $\longrightarrow$ Bool be two decision problems in **TotCompFunc**. We say that f is *polynomial reducible* to g or *f polynomially reduces* to g if there is an h: Seq $\longrightarrow$ Seq$'$ in **DTIME**(Poly)

such that

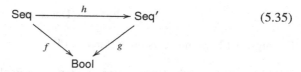

$$(5.35)$$

commutes. We call h the polynomial reduction, written as $f \leq_p g$. If we further have that $g \leq_p f$ then we write $f \equiv_p g$ and say f and g are in the same *complexity class*.

To form the category of decision problems and polynomial reductions, consider the functors

$$\textbf{DTIME}(\text{Poly}) \xhookrightarrow{\text{Inc}} \textbf{TotCompFunc} \xleftarrow{\text{Const}_{\text{Bool}}} \textbf{1} \qquad (5.36)$$

where the left functor is an inclusion functor and $\text{Const}_{\text{Bool}}$ takes the single object in **1** to the type Bool. Now the category of decision problems is the comma category **Decision** $= (\text{Inc}, \text{Const}_{\text{Bool}})$. The objects of this category are total computable decision problems and the morphisms are polynomial reductions from one decision problem to another.

There are two subcategories of **TotCompFunc** of interest: **DTIME**(Poly) and **NTIME**(Poly). These are all deterministic polynomial computable functions, and all nondeterminisitic polynomial functions, respectively. They sit in the diagram

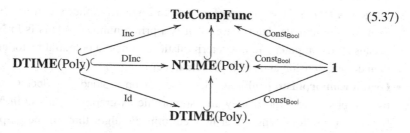

$$(5.37)$$

Definition 5.11. The comma category $(\text{DInc}, \text{Const}_{\text{Bool}})$ which consists of nondeterministic polynomial decision problems and (deterministic) polynomial reductions is called the complexity class **NP**. (Perhaps a better name would have been **NPTIME**, but we follow the convention.) The comma category $(\text{Id}, \text{Const}_{\text{Bool}})$ which consists of deterministic polynomial decision problems and (deterministic) polynomial reductions is called the complexity class **P**. (Again, a better name would be **DPTIME**.) By Exercise 2.10, the inclusion

$$\textbf{DTIME}(\text{Poly}) \hookrightarrow \textbf{NTIME}(\text{Poly}) \hookrightarrow \textbf{TotCompFunc} \qquad (5.38)$$

induces the inclusion

$$\mathbf{P} \hookrightarrow \mathbf{NP} \hookrightarrow \mathbf{Decision}. \tag{5.39}$$

Example 5.12. Some decision problems in **P**.

- **Path problem.** Path: String × String × String ⟶ Bool. Given a graph G with two vertices, s and t, determine if there is a path in the graph from s to t. (This is in **P** because a computer can start at s and perform a systematic (depth-first) search through the graph to find t. If there are n nodes there are at most n^2 edges. Therefore the decision problem is solvable in quadratic time.)

- **Euler cycle problem.** Euler: String ⟶ Bool. Given a graph, determine if there is a path that starts and finishes at the same vertex and crosses every edge exactly once. Such a path is called an "Euler cycle." (It can be shown to be in **P** because of a trick that Leonhard Euler taught us when he solved the Königsberg bridge problem. To determine if a graph has an Euler cycle, make sure there is an even number of edges coming out of each vertex. A graph with n vertices has at most n^2 edges so the algorithm is n^2.)

- **Prime problem.** Prime: Nat ⟶ Bool. Given a positive whole number, determine if it prime. (The fact that this problem is in **P** was a major research in Agrawal et al., 2002.)

Example 5.13. Some decision problems in **NP**.

- **Factoring decision problem.** Factor: Nat×Nat ⟶ Bool. Given two positive integers $m < n$, is there a factor of n between 2 and m? Notice that if $m = \sqrt{n}$ then we are actually determining if n is a prime number. (This is in **NP** because the computer can nondeterministically guess a potential factor and then do one division and see if the remainder is zero.)

- **Graph isomorphism problem.** GraphIso: String × String ⟶ Bool. Given two graphs, determine if they are isomorphic as graphs. (This is in **NP** because of the following linear nondeterministic algorithm. If the graphs have a different number of vertices, then they are clearly not isomorphic. On the other hand, if they have the same number of vertices, say n, then take one graph and nondeterministically look at all $n!$ permutations and see if any permutation maps isomorphically to the second graph. A permutation π is an isomorphism of graphs when there is an edge between x and y if and only if there is an edge between $\pi(x)$ and $\pi(y)$.)

- **Discrete logarithm problem.** DiscLog: Nat × Int × Int × Nat ⟶ Bool. First some preliminaries. Let us say that we are given real numbers b and x such that $x = b^y$ for an unknown y. Determining y is not hard. You solve the problem by taking the logarithm based at b of x. That is, $\log_b(x) = \log_b(b^y) = y$.

This is not computationally difficult because exponentiation is continuous and monotonic and hence its inverse, the logarithm function, is also continuous and monotonic. The discrete logarithm problem is the same problem but rather then dealing with the real numbers, it deals with $\mathbb{Z}_p$ for some prime p. (The definition can actually be done with any group.) Here the exponentiation is neither continuous nor monotonic. For example if $p = 17$ and $b = 5$ then we have the following values of modular exponentiation:

y	1	2	3	4	5	6	7	$\cdots$
$x = 5^y$ Mod 17	5	8	6	13	14	2	10	$\cdots$

As you can see there is no clear pattern in x.

Let us state the discrete logarithm problem as a decision problem. Given a prime p, a generator $b \in \mathbb{Z}_p$, an integer $0 < c < p$, and an upper bound $d \in \mathbb{N}$, determine whether there exists a $1 \le y \le d$ such that $b^y \equiv c$ Mod p. (The only known way to solve the discrete logarithm problem is to nondeterministically try all the possible exponents below the bound. This means it is in **NP**.)

We will meet the Factoring decision problem and the Discrete logarithm problem during our discussion of cryptography in §2 of the supplement to this Element.

Example 5.14. Here are some more problems in **NP**. We will see that these are a different type of **NP** problems from the problems in Example 5.13.

- **Satisfiability problem.** SAT: String $\longrightarrow$ Bool. Given a logical formula, determine if there is a way to assign values to the variables such that the formula is true, i.e., tell if the formula is not a contradiction. Another way to say this is tell if there is a "T" in the final column of the truth table for the formula. (This is in **NP** because one can nondeterministicly try all 2^n rows of the truth table where n is the number of variables in the formula. For each row, evaluate the formulas with those truth values and see if any result is true.)
- **Knapsack problem.** Knapsack: Nat*$\times$Nat*$\times$Nat$\times$Nat $\longrightarrow$ Bool. The inputs to the problem are n objects, each with a size and a price, a size capacity, and a goal profit. Determine if there is a subset of elements that will fit in the knapsack and whose sum is more than or equal to goal profit. Technically, given $(s_1, s_2, \ldots, s_n)$, $(p_1, p_2, \ldots, p_n)$, a capacity C, and goal G, determine if there is an $X \subseteq \{1, 2, \ldots, n\}$ such that

$$\sum_{x \in X} s_x \le C \quad \text{and} \quad \sum_{x \in X} p_x \ge G. \quad (5.40)$$

(This is in **NP** because the computer can nondeterministically try all the 2^n subsets. For each subset, see if everything is less than the capacity and if the sum of the profits is more than the goal. If any of the subsets satisfy these requirements, respond true.)

- **Hamiltonian cycle problem.** Hamiltonian: String $\longrightarrow$ Bool. Given a graph, determine if there exists a cycle (a path which starts and finishes at the same vertex) that traverses every vertex exactly once. (This is in **NP** because a computer can nondeterministically try all $n!$ possible routes for a graph of size n. For each possible route, make sure that there is an edge between the pairs of vertices of that possible route. If there is, then respond true.)

- **Traveling salesman problem.** TSP: String $\times$ Nat $\longrightarrow$ Bool. Given a collection of cities that a traveling salesman wants to visit and a budgetary limitation, determine if there is a route that he can take that is within his budget. More formally, given a weighted, complete graph with a number K, determine if there exists a cycle that traverses every vertex exactly once, and the sum of weights of the edges used is less than or equal K. (This is also in **NP** because, again, one can nondeterministically try all possible routes.)

- **Subset sum problem.** SSP: Nat* $\times$ Nat $\longrightarrow$ Bool. Given a set of natural numbers, and a number C, determine if there is a subset of the numbers that sums to C. More formally, given $(s_1, s_2, \ldots, s_n)$ and a C, determine if there is an $X \subseteq \{1, 2, \ldots, n\}$ such that

$$\sum_{x \in X} s_x = C. \tag{5.41}$$

(This is in **NP** because one can nondeterministically try all 2^n subsets and see if any of them sum to C.)

- **Set partition problem.** SPP: Nat* $\longrightarrow$ Bool. Given a set of numbers, determine if there is a way to split the numbers into two sets such that the sum of one subset is equal to the sum of the other subset. More formally, given $(s_1, s_2, \ldots, s_n)$, determine if there is a $X \subseteq \{1, 2, \ldots, n\}$ such that

$$\sum_{x \in X} s_x = \sum_{x \notin X} s_x. \tag{5.42}$$

(This too is in **NP** because one can nondeterministically try all subsets.)

Notice that all of these problems do not seem to have a simple deterministic polynomial time algorithm.

Advanced Topic 5.15. Learning that these problems are essentially unsolvable for large inputs is interesting. However, there is another, more practical branch of complexity theory that is of interest. While we probably cannot get the best answer to these problems, we might be able to get to a close approximation to the right answer. *Approximation algorithms* are polynomial algorithms for hard problems that give close solutions. Researches spend time classifying how complicated these algorithms are and how close of a solution they usually find. There are chapters about approximation algorithms in textbooks (Papadimitriou, 1994; Arora and Barak, 2009; Sipser, 2006).

We are not only interested in the objects of **P** and **NP**. We are also interested in the reductions between such problems. The notion of polynomial reduction is very important. In Diagram (5.35), since h is in **DTIME**(Poly), if g is also in **DTIME**(Poly), then by composition, so is f. That is, if g is in **P**, then f is in **P**. The contrapositive of this statement is more interesting: If f is not in **DTIME**(Poly), then neither is g. That is, if f is not in **P**, then neither is g in **P**.

Example 5.16. Probably the simplest example of a reduction of one problem to another is the reduction from the Set partition problem (SPP) to the Subset sum problem (SSP).

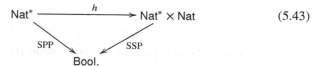

$$ \text{(5.43)} $$

Given an input for the SPP such as $\{23, 27, 3, 34, 7, 31, 11\}$ we can make it into an input to the SSP by taking the same numbers and choosing C as half of the sum of all the elements. In this case, we form the input to the SSP as $\{23, 27, 3, 34, 7, 31, 11\}$ and $C = (\sum_{i=1}^{n} x_i)/2 = 136/2 = 68$. With this input we can see that there is a solution to the SSP using the subset $\{7, 27, 34\}$. With this solution, we can formulate a solution to the SPP by taking this subset of numbers and the subset of numbers not in it, i.e., $\{7, 27, 34\}$ and $\{3, 11, 31, 23\}$. The function h that takes the input from one problem to the other problem is clearly polynomial.

The most important part of the reduction is that if there is a polynomial algorithm for solving the SSP, then using function h, we can find a polynomial algorithm for solving the SPP. Equivalently, if there is no polynomial time algorithm to solve the SPP, then there is definitely no polynomial time algorithm to solve the SSP. Both ways of saying it mean the same thing: the SSP is as hard or harder than the SPP.

Example 5.17. Another example of a polynomial reduction problem from one **NP** problem to another is the reduction of the Hamiltonian cycle problem to the Traveling salesman problem. We describe a deterministic polynomial map h

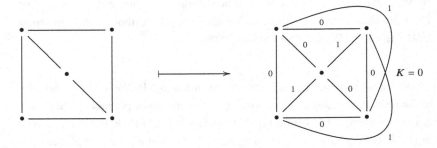

Figure 5.12 A reduction from an instance of the Hamiltonian cycle problem to an instance of the Traveling salesman problem

such that the following triangle commutes:

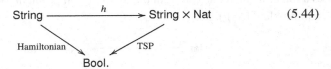

$$(5.44)$$

The map h has three tasks to perform with the graph: (i) Complete the graph; that is, add in all the edges needed so that there is an edge from every vertex to every other vertex. (ii) Add weights to the graph. We do this by assigning all the edges from the original graph the number 0 and all the new edges – added to make the graph complete – the weight 1. (iii) Set the budget $K = 0$. This means that we want a route that has zero weight. Another way of saying this is that we can only use the edges from the original graph and not the new added edges. Figure 5.12 is an example of such a reduction.

Notice that there is a successful Traveling salesman route if and only if there is a Hamiltonian graph. We have shown that if one can solve the Traveling salesman problem in polynomial time, then one can definitely solve the Hamiltonian graph problem. The TSP is as hard or harder than the Hamiltonian cycle problem.

Exercise 5.18. Show that there is a polynomial reduction from the Subset sum problem to the Knapsack problem.

We have shown that since there is a morphism in **NP** from the SSP to the SSP, then the SSP is as hard or harder than the SPP. From another morphism in **NP** we showed that the Traveling salesman problem is as hard or harder than the Hamiltonian path problem. Exercise 5.18 asks for a morphism in **NP** to show that the Knapsack problem is as hard or harder than the SSP. To discuss the hardest **NP** problems we need the notion of a weak terminal object in **NP**.

Definition 5.19. A terminal object t in a category is an object such that for any object a there is *exactly one* morphism $a \longrightarrow t$. Define a *weak terminal object* in a category to be an object w such that for every object a in the category there is *at least one* morphism $a \longrightarrow w$.

The problems in **NP** that are weak terminal objects are the hardest problems in **NP** and are called *NP-complete problems*. The full subcategory of **NP** of all NP-complete problems is denoted **NPComplete**. These are the nondeterministic polynomial decision problems such that every nondeterministic polynomial decision problem polynomially reduces to one of them. We have the inclusion of categories **NPComplete** $\hookrightarrow$ **NP**.

Given a weak terminal object w, any map $h: w \longrightarrow w'$ ensures that the object w' is also a weak terminal object. In terms of NP-complete problems, this means that if f is an NP-complete problem, and there exists a polynomial reduction from f to NP-problem g, then g is also an NP-complete problem. So to find a cadre of NP-complete problems, we have to find a single NP-complete problem first. Once we have such a problem, any object with a morphism from that object will also be NP-complete. Logic gives us this first example.

Theorem 5.20 (The Cook–Levin Theorem). SAT: *String* $\longrightarrow$ *Bool is a weak terminal object in* **NP**. *That is,* SAT *is an NP-complete problem.*

Proof. Let us emphasize what this means. For SAT to be NP-complete then *every* NP problem reduces to it. Over the past several decades, researchers have described thousands of **NP** problems. There are still thousands more to be described in the future. How are we to show that everyone of these problems can be reduced to SAT?

We have to show that for any $g:$ Seq $\longrightarrow$ Bool in **NP** there is a polynomial reduction $h_g:$ Seq $\longrightarrow$ String such that

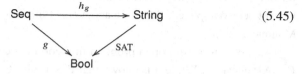

$$\text{(5.45)}$$

commutes. The one thing we know about every problem in **NP** is that – by definition – there is a nondeterministic Turing machine that can execute the function. Let T_g be a nondeterministic total Turing machine that computes g. Consider the functor

$$\textbf{TotTuring} \xrightarrow{\ L_t\ } \textbf{Logic}. \qquad (5.46)$$

Now $L_t(T_g)$ is a family of sequences of logical formula that describes the workings of T_g. If x is an input to T_g, then $L_t(T_g)[x]$ is a sequence of logical formulas whose conjunction (and) is either satisfiable or not satisfiable depending on whether or not there is a successful computation of T_g with input x. We define $h_g(x) = L_t(T_g)[x]$. The point of the construction is that $g(x)$ is true if and only if $L_t(g)[x]$ is satisfiable. The hard part of the proof is to show that h_g is polynomial. One can find that part of the proof in Section 2.6 of Garey and Johnson (1979), Section 34.3 of Cormen et al. (2009), and Section 7.4 of Sipser (2006).   □

With SAT established as an NP-complete problem, we can find other NP-complete problems by finding polynomial reductions to those other problems. Researchers have formed a whole web of reductions: see Examples 5.16, 5.17 and Exercise 5.18 for instance. There are now literally thousands of NP-complete problems, including every one mentioned in Example 5.14.

On the other hand, the NP problems in Example 5.13 are not known to be NP-complete. They are hard problems, but at the moment no one knows any reduction from an NP-complete problem to any of them. In fact, the three problems mentioned in Example 5.13 are some of the few that are tentatively called "NP-intermediate." For each such problem, one of three things can happen: (i) someone can find a polynomial algorithm for the problem and it will be in **P**, or (ii) someone can find a reduction from a problem in **NPComplete** to the problem and then it will be in **NPComplete** or (iii) they can remain NP-intermediate. At the moment we do not know.

In 1994, Peter Shor described a polynomial algorithm for a quantum computer to solve the factoring decision problem (Shor, 1994). There are no known polynomial algorithms for quantum computers to solve any NP-complete problems. Many researchers inferred that polynomial algorithms for quantum computers were made for this NP-intermediate class, and began a huge search for one to solve the graph isomorphism problem. As far as I know, this task has not been completed. See more about quantum computing in Yanofsky and Mannucci (2008).

The most prominent open problem in theoretical computer science is the **P** = **NP** question. While it is known that **P** is a subcategory of **NP**, it remains an open question to tell if these categories are really the same category. In other words, is every problem in **NP** also in **P** and hence **P** = **NP** or is there a problem in **NP** that is not in **P** and hence **P** ≠ **NP**. Alas, this question will not be answered in these pages.

NP-complete problems are central to the **P** = **NP** question. If one shows that any NP-complete problem can be decided by a polynomial Turing machine then all the morphisms in **NP** can be shown to be in **P** and **P** = **NP**. That is, if there is any morphism in **NP** from an object in the subcategory **NPComplete** to any object in the subcategory **P**, then **P** = **NP**. In contrast, if we can find one problem in **NP** that does not have a polynomial Turing machine, then we can show that **P** ≠ **NP**. From the fact that this problem has been around for half a century and no one has found a polynomial algorithm for the myriad of NP-complete problems, many believe that **P** ≠ **NP**. Others are less convinced.

A complexity class of decision problems is a slice category of the form **C** = **A**/Bool. From Exercise 2.4 we know that the total computable function NOT: Bool ⟶ Bool induces another complexity class which is called co**C**.

Example 5.21. Some decision problems in co**P**.

- **coPath problem.** coPath: String × String × String ⟶ Bool. Given a graph and two points, s and t, determine if there is no path from s to t in the graph. (One needs to traverse through all vertices starting from s and if you visit all the vertices that you can and did not see t, then answer true.)
- **coEuler cycle problem.** coEuler: String ⟶ Bool. Given a graph, determine if there does not exist a cycle through all the vertices that hits every edge exactly once. (One needs to go through all vertices and determine if any vertex has an odd number of edges number of edges coming out of it. If it does, then answer true.)
- **coPrime problem.** coPrime: Nat ⟶ Bool. Given a natural number, determine if the number is not a prime.

It should be obvious that all these problems can be determined in polynomial time. More generally, it is true that **P** = co**P**.

Example 5.22. Some decision problems in co**NP**.

- **unSatisfiable problem.** unSatisfiable: String ⟶ Bool. Given a Boolean formula, determine if it is impossible to satisfy the formula. (In contrast to Satisfiability, here one seems to have to look at the entire truth table and make sure that all the final entries are false.)
- **unHamiltonian problem.** unHamiltonian: String ⟶ Bool. Given a graph, determine if there is no Hamiltonian cycle. (This also seems harder than Hamiltonian cycle problem because one has to try all possible routes and make sure none of them are Hamiltonian. We cannot stop when we find something. We have to check all.)
- **unSubset sum problem.** unSSP: Nat* × Nat ⟶ Bool. Given a list of finite whole numbers and a capacity C, make sure that no subset of numbers adds up to C. (This seems harder than the Subset sum problem because you have to try all subsets.)

All three of these problems seem harder than **NP**. It remains an open question if **NP** = co**NP**.

Advanced Topic 5.23. We know that factoring, graph isomorphisms, and discrete logarithms are NP-problems which have not been shown to be complete. We are also not sure that they are NP-intermediate. They might be in **P** or be shown to be in NP-Complete. Question: are there, in fact, any NP-intermediate problems? In 1975, Richard E. Ladner proved a theorem that says if **P** is not equal to **NP**, then there is a class of NP-intermediate problems $\textbf{P} \hookrightarrow \text{Intermediate} \hookrightarrow \textbf{NP}$. In fact he showed that if $\textbf{P} \neq \textbf{NP}$ there are an infinite number of complexity classes between them; see Section 3.3 of Arora and Barak (2009). Researchers have gone on to prove a generalized Ladner theorem that says between any two complexity classes of a certain type, there are infinitely many intermediate classes. This is similar to the Friedberg–Muchnik theorem result mentioned in Advanced Topic 4.22.

Research Project 5.24. This will be more in the way of a dream than a research project. First some preliminaries. To every decision problem there is usually associated a "search space." For example, with the factoring decision problem, the search space is the set of all numbers less then the square root of the input. One of these numbers might be a factor of the input. For the Traveling salesman problem, the set of all possible routes (cycles) that hit every city is the search space. For the Knapsack problem, the set of all possible subsets of objects is the search space. When a computer solves a decision problem, it searches through the search space. Many times the search space is small and the computer can easily find the solution. Other times, the search space is large, but there are tricks to find the solution with ease. And still other times, the search space is large and only a lengthy brute-force search will succeed.

There is an intuition about the difference between problems in **P** and problems in **NP** that goes as follows: problems in **P** have well-structured search spaces. Programs that solve problems in **P** use this structure to find solutions with fewer operations. In contrast, problems in **NP** have search spaces that do not have a lot of structure to exploit. It should be noted that even for **NP** problems, the search space usually has a tree structure. However, this is not good enough for programs for finding the solution quickly.

It would be wonderful if this intuition was formalized into clear statements. This is where category theory comes in. Over the past few decades, category theorists have classified and categorized many different ways of describing structures. They have invented hierarchies of algebraic theories, monads, operads, etc. It would be nice to classify what type of such structure-defining devices produce search spaces that are in **P** as opposed to **NP**. This will give a purely syntactical way of describing what is in **P** and what is in **NP**.

Another project that might be of interest is to look at the logic of decision problems. Consider the computable, Boolean logical conjunction operation $\wedge:$ Bool $\times$ Bool $\longrightarrow$ Bool. This induces a functor

$$\wedge: \textbf{Decision} \times \textbf{Decision} \longrightarrow \textbf{Decision} \qquad (5.47)$$

by taking $f:$ Seq $\longrightarrow$ Bool and $g:$ Seq$'$ $\longrightarrow$ Bool to

$$f \wedge g = \wedge \circ (f \otimes g): \text{Seq} \times \text{Seq}' \longrightarrow \text{Bool} \times \text{Bool} \longrightarrow \text{Bool}. \qquad (5.48)$$

This functor respects the morphisms in **Decision** because if $f \leq_p f'$ and $g \leq_p g'$ then $f \wedge g \leq_p f' \wedge g'$. There is a similar functor for the computable logical disjunction ($\vee$).

We can also talk about quantifiers. For every type Seq$'$ there is a silly decision problem $s:$ Seq$'$ $\longrightarrow$ Bool that takes everything to 1. There is also a projection function $\pi_1:$ Bool $\times$ Bool $\longrightarrow$ Bool that projects the first value and ignores the second value. We can use these to show any decision problem $f:$ Seq$'$ $\longrightarrow$ Bool can trivially be thought of as a decision problem from Seq $\times$ Seq$'$. In detail,

$$f: \text{Seq} \longrightarrow \text{Bool} \quad \mapsto \quad \pi_1 \circ (f \times s): \text{Seq} \times \text{Seq}' \longrightarrow \text{Bool} \times \text{Bool} \longrightarrow \text{Bool}.$$

This means we are just using the silly decision problem as a dummy variable. If we look at the subcategory of all the decision problems with source Seq, and call this subcategory **Decision(Seq)** then we just defined an inclusion functor

$$\text{Inc}: \textbf{Decision(Seq)} \hookrightarrow \textbf{Decision(Seq} \times \textbf{Seq}'). \qquad (5.49)$$

If there is a left adjoint to this inclusion functor, then it is the existential quantifier, i.e.,

$$\exists_{\text{Seq,Seq}'}: \textbf{Decision(Seq} \times \textbf{Seq}') \longrightarrow \textbf{Decision(Seq)}. \qquad (5.50)$$

If a right adjoint exists, then it is universal quantifier (i.e., $\forall_{\text{Seq,Seq}'}$.)

How much of this logic can be developed? This will be intimately related to descriptive complexity theory and finite model theory mentioned in Advanced Topic 3.59.

5.3 Space Complexity

Till now we have concentrated on the time resource. Let us look at how much space a computer can use. We will begin by examining problems that demand little space. When we dealt with time, we did not discuss problems that demand less less time than the size of the input (i.e., sublinear time) because a Turing machine will not be able to see the entire input in sublinear time. However, we can talk of problems that demand sublinear space in their work tape.

Remember that **DSPACE**(Log) are those computable functions that can be computed using $\log n$ space or less. Such functions are called *log-space computable*.

Definition 5.25. In analogy with what we did when defining **NP** and **P** in Diagram (5.37), let us consider two subcategories of **TotCompFunc**

$$\textbf{DSPACE}(\text{Log}) \hookrightarrow \textbf{NSPACE}(\text{Log}) \hookrightarrow \textbf{TotCompFunc}. \qquad (5.51)$$

We use these categories to form two comma categories that are complexity classes: (i) **L** whose objects are deterministic logarithm space decision problems and whose morphisms are log-space transducers; and (ii) **NL** whose objects are nondeterministic, logarithmic space, decision problems and whose morphisms are log-space transducers. Weak terminal objects in these categories are called *L-complete* and *NL-complete* problems respectively.

In 1987, Neil Immerman and Róbert Szelepcsényi independently proved the Immerman–Szelepcsényi theorem which says that **NL** = co**NL**. This is in stark contrast to **NP** and co**NP** which are thought (but not proven) to be different.

Now let us go on from the subset of Log functions and talk of the subset of Poly functions.

Definition 5.26. There are inclusions

$$\textbf{DSPACE}(\text{Poly}) \hookrightarrow \textbf{NSPACE}(\text{Poly}) \hookrightarrow \textbf{TotCompFunc}. \qquad (5.52)$$

The first two categories are subcategories of **TotCompFunc** that correspond to total computable functions that can be computed using a polynomial amount of space deterministically and nondeterministically, respectively. Using diagrams analogous to Diagram (5.37), we can form the category of decision problems **PSPACE** and **NPSPACE**. Weak terminal objects in these categories are *PSPACE-complete* and *NPSPACE-complete* problems respectively.

Let us give one of the main results about the space resource. When dealing with time complexity, the big open question is the relationship between **P** and **NP**. By contrast, the analogous question for space complexity was answered by Walter Savitch in 1970.

Theorem 5.27 (Savitch's Theorem). *The inclusion functor* **PSPACE** $\hookrightarrow$ **NPSPACE** *is actually the identity. That is,* **PSPACE** = **NPSPACE**.

Proof. The proof basically shows that every nondeterminstic computation that uses $f(n)$ spaces can be imitated by a deterministic computation that uses $f(n)^2$ spaces. In particular, if $f(n)$ is a polynomial, then $f(n)^2$ is also a polynomial. We conclude that all the nondeterministic computable decision problems in **NPSPACE** are also in **PSPACE**. The proof uses the fact that space can be reused. That is, one can deterministically imitate the nondeterministic function recursively, and reuse the space that was used and no longer needed. This is in stark contrast to the time resource which cannot be reused. The details of the proof can be found in Section 8.1 of Sipser (2006), Section 7.3 of Papadimitriou (1994), and Section 4.2 of Arora and Barak (2009).   □

One can also talk about decision problems that demand deterministic and nondeterministic exponential space. They form categories called **DEXPSPACE** and **NEXPSPACE**. There are examples of such such problems and there is a notion of completeness in those categories. We can elaborate, but we are out of . . . space.

All the complexity classes of decision problems that we saw are summarized as

$$\mathbf{L} \to \mathbf{NL} = co\mathbf{NL} \to \mathbf{P} = co\mathbf{P} \overset{\mathbf{NP}}{\underset{co\mathbf{NP}}{\rightrightarrows}} \mathbf{PSPACE} \to \mathbf{DEXPSPACE}$$

where the arrows are inclusion functors.

Research Project 5.28. How would one talk about probabilistic computation? Our entire discussion has been about machines that exactly compute functions. How about machines that *almost* compute functions? If we can formulate this, then we can talk about the probablistic complexity classes: ZPP (zero-error probabilistic polynomial time), BPP (bounded-error probabilistic polynomial time), RP (randomized polynomial time), etc. This will also be very important in order to talk about the quantum complexity class: BQP (bounded-error quantum polynomial time).

Further Reading

Lewis and Papadimitriou (1997); Papadimitriou (1994); Arora and Barak (2009) are popular textbooks in complexity theory. Chapters 7–10 of Sipser (2006) are a great introduction to complexity theory. See also Chapters 27–30 of Rich (2008), and Chapter 12 and 13 of Hopcroft and Ullman (1979). Dean (2016) is a fascinating article on complexity theory from a philosophical perspective. For a nice popular science overview of the field see Chapter 5 of

Yanofsky (2013).

For particular topics:

- Measuring complexity: Chapters 1–4 of Cormen et al. (2009), Chapters 4 and 7 of Sipser (2006).
- NP-complete problems: Garey and Johnson (1979), and Chapter 34 of Cormen et al. (2009).
- Space complexity: Chapter 8 of Sipser (2006), Chapter 19 of Papadimitriou (1994), and Chapter 29 of Rich (2008).

6 Diagonal Arguments

Many systems have the ability to describe something about themselves or have self reference. An English sentence has the ability (not only to describe something about the universe, but also) to describe something of itself. "This sentence has five words" is saying something true about itself. "This sentence does not have seven words" states something false about itself. Language has self reference. As we wrote in the introduction, Cantor showed that sets can talk about themselves, and Gödel showed that mathematical statements can talk about themselves. Alan Turing continued in this tradition and showed that computers can have self reference. These diverse areas all have self reference which causes paradoxical situations. It is remarkable that there is a common scheme of many self-referential paradoxes. When objects in a system have the ability to negate some basic aspect of themselves, there will be a paradox. This scheme seems to be a fundamental facet of all reasonings.

In theoretical computer science and mathematics these self-referential phenomena take the form of a diagonal argument. There are many theorems that are proven using these arguments. Even though these ideas arise in many diverse areas of theoretical computer science and mathematics, it turns out that all of them can be seen as a consequence of a simple theorem of category theory. This is yet another demonstration of the power of category theory.

6.1 Cantor's Theorem

Historically diagonal arguments go back to the German mathematician Georg Cantor in the last decades of the nineteenth century. He was the first to prove that there were different levels of infinity. For pedagogical purposes, we will state the theorem and examine the proof in three levels of increasing abstraction. By doing this, we will distill its essence and get a categorical theorem and proof. We then move on to show all the diverse instances of the theorem.

Theorem 6.1. *There does not exist an onto function from the set of natural numbers,* $\mathbb{N}$, *to the powerset of natural numbers,* $\mathcal{P}(\mathbb{N})$. *That is,* $\mathbb{N}$ *is a smaller set than* $\mathcal{P}(\mathbb{N})$.

Proof. First, the standard proof. Cantor proved that there does not exist an onto function $f: \mathbb{N} \longrightarrow \mathcal{P}(\mathbb{N})$ by using a proof by contradiction. The contradiction will involve finding a subset of natural numbers $D_f \in \mathcal{P}(\mathbb{N})$ that is not in the image of f and hence f is not onto. Let us motivate the definition of D_f. The set must satisfy the following property

$$D_f \text{ is not in the image of } f.$$

Or, more precisely,

$$D_f \text{ is different from every set } f(n) \text{ for all } n.$$

How should D_f be different? What number should be in $f(n)$ while not be in D_f or vice versa? Cantor answered this in a systematic way.

$$D_f \text{ is different from every set } f(n) \text{ for all } n \text{ because } D_f \text{ and } f(n) \text{ are different}$$
$$\text{at } n.$$

This means that $n \in D_f \iff n \notin f(n)$. In symbols this becomes

$$D_f = \{n \in \mathbb{N} : n \notin f(n)\}. \tag{6.1}$$

Now we go on to formally prove that D_f is not in the image of f. If D_f was in the image of f there would be an $n_0 \in \mathbb{N}$ such that $f(n_0) = D_f$. Let us ask a simple question: is n_0 in D_f or not?

$$n_0 \in D_f \qquad \underbrace{\Longleftrightarrow}_{\text{bec. } D_f = f(n_0)} \qquad n_0 \in f(n_0) \qquad \underbrace{\Longleftrightarrow}_{\text{bec. of def of } D_f} \qquad n_0 \notin D_f.$$

This is a contradiction. It must be that our assumption that D_f was in the image of f was wrong. $\qquad\qquad\qquad\qquad\qquad\qquad\qquad\qquad\qquad\qquad\qquad\qquad\square$

Now let us look at the same proof from a more categorical perspective.

Proof. As is well known, a subset of a set $\mathbb{N}$ can be seen as a characteristic function $\chi: \mathbb{N} \longrightarrow 2$ where the set $2 = \{0, 1\}$. This means that $\mathcal{P}(\mathbb{N})$ is essentially the same as $\text{Hom}_{\mathbf{Set}}(\mathbb{N}, 2) = 2^{\mathbb{N}}$. Rather than considering functions $f: \mathbb{N} \longrightarrow \mathcal{P}(\mathbb{N})$, we consider functions $f: \mathbb{N} \longrightarrow 2^{\mathbb{N}}$. This means that for every $n \in \mathbb{N}$ we have a characteristic function $f(n): \mathbb{N} \longrightarrow 2$. Cantor's theorem says there is no onto f or that for every $f: \mathbb{N} \longrightarrow 2^{\mathbb{N}}$, there is some characteristic function $D_f: \mathbb{N} \longrightarrow 2$ which is not in the image of f. By examining Equation (6.1) we define D_f as

$$D_f(n) = \begin{cases} 1: & \text{if } f(n)(n) = 0 \\ 0: & \text{if } f(n)(n) = 1 \end{cases} \qquad (6.2)$$

Formally, if D_f was in the image of f then there would be an $n_0 \in \mathbb{N}$ such that $f(n_0) = D_f$, and, similar to what we did above, we would have

$$D_f(n_0) = 1 \quad \Longleftrightarrow \quad f(n_0)(n_0) = 1 \quad \Longleftrightarrow \quad D_f(n_0) = 0. \quad (6.3)$$

This is clearly a contradiction.   □

Let us look at this proof from an even more categorical perspective. Here we use the fact that the category of sets is Cartesian closed. This means that set functions of the form $A \longrightarrow C^B$ are equivalent to set functions of the form $A \times B \longrightarrow C$.

Proof. Rather than looking at set functions of the form $\mathbb{N} \longrightarrow 2^{\mathbb{N}}$, let us examine functions of the form $f: \mathbb{N} \times \mathbb{N} \longrightarrow 2$. Such a function can be viewed as a two-dimensional infinite array with Boolean entries as can be seen in Figure 6.13. Every row of this matrix represents a set of numbers. The nth row represents the set $f(n, -)$ and entries in the row tell which numbers are in that set. $f(n, m) = 1$ means that $m \in f(n, -)$ and $f(n, m) = 0$ means $m \notin f(n, -)$. Cantor's theorem then says there exists a set of numbers, described by $D_f: \mathbb{N} \longrightarrow 2$ that is not in the listing of all the sets described by f. We can easily form the set D_f by the composition of the following maps in the category of sets.

f	0	1	2	3	4	5	$\cdots$
0:	1	1	0	1	1	1	$\cdots$
1:	1	1	1	1	0	0	$\cdots$
2:	0	0	0	0	0	0	$\cdots$
3:	0	1	0	1	0	0	$\cdots$
4:	1	1	1	1	1	1	$\cdots$
5:	0	1	1	0	1	0	$\cdots$
$\vdots$	$\vdots$		$\vdots$		$\vdots$		

Figure 6.13 A function $f: \mathbb{N} \times \mathbb{N} \longrightarrow 2$ purported to describe all subsets of $\mathbb{N}$

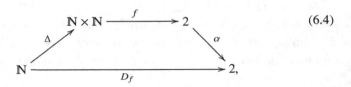

$$(6.4)$$

where $\Delta: \mathbb{N} \longrightarrow \mathbb{N} \times \mathbb{N}$ is defined as $\Delta(n) = (n,n)$ and $\alpha: 2 \longrightarrow 2$ is defined as $\alpha(1) = 0$ and $\alpha(0) = 1$. Using these maps, we can see that $D_f(n) = \alpha(f(n,n))$ which is exactly what Equation (6.2) describes. In terms of Figure 6.13, D_f can be seen as the changed diagonal elements of the infinite matrix as in Figure 6.14. We change the (n,n) element for each n. This is where we get the name "diagonal argument."

f	0	1	2	3	4	5	$\cdots$
0:	$\alpha(1) = 0$	1	0	1	1	1	$\cdots$
1:	1	$\alpha(1) = 0$	1	1	0	0	$\cdots$
2:	0	0	$\alpha(0) = 1$	0	0	0	$\cdots$
3:	0	1	0	$\alpha(1) = 0$	0	0	$\cdots$
4:	1	1	1	1	$\alpha(1) = 0$	1	$\cdots$
5:	0	1	1	0	1	$\alpha(0) = 1$	$\cdots$
$\vdots$	$\vdots$		$\vdots$		$\vdots$		

Figure 6.14 A function $f: \mathbb{N} \times \mathbb{N} \longrightarrow 2$ that is purported to describe all subsets of $\mathbb{N}$, and diagonal function $D_f: \mathbb{N} \longrightarrow 2$ that is different than every row of f.

From here it is obvious that the function D_f is different from every row of the matrix; that is, $D_f(-) \neq f(n,-)$ for all $n \in \mathbb{N}$. They are different because

$$D_f(n) = \alpha(f(n,n)) \neq f(n,n). \tag{6.5}$$

This means that D_f describes a subset of natural numbers that is not on the list of all subsets described by f. We say D_f "diagonalizes out" of the subsets described by f. □

Now let us distill what was used in this theorem and proof about sets to get a theorem and proof about categories. Cantor discussed sets and functions, but any category theorist worth her weight in salt will abstract this to talk about any category **A** with certain properties. What properties did we use? We needed a product $\times$ and a terminal object 1. Rather than talking about a set function $f: \mathbb{N} \times \mathbb{N} \longrightarrow 2$ we should talk about a morphism in **A** of the form $f: a \times a \longrightarrow y$ for some objects a and y in **A**. (This is the core of self reference. We are seeing how elements of a are dealing with elements of **A**.)

The function α helped D_f be different from every row by changing every 0 to 1 and every 1 to 0. If we are dealing with a set other than $\{0,1\}$ we would need a function that changes every element to some other element. Formally, if we were dealing with a set Y, then we would need a function $\alpha: Y \longrightarrow Y$ such that for all $x \in Y$, we have $\alpha(x) \neq x$. Let us formally define the opposite of this.

Definition 6.2. First a simple definition in **Set**. Consider a set Y and a set function $\alpha: Y \longrightarrow Y$. We call $x \in Y$ a *fixed point* of α if $\alpha(x) = x$. That is, the output is the same (or fixed) as the input. We can write the element x by talking about a function $p_x: \{*\} \longrightarrow Y$ such that $p_x(*) = x$. Saying that x is a fixed point of α amounts to saying that $\alpha \circ p_x = p_x$, i.e., the following diagram commutes:

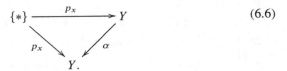

$$(6.6)$$

Let us generalize this to any category **A** with a terminal object 1. Let y be an object in **A** and $\alpha: y \longrightarrow y$ be a morphism in **A**. Then we say $p: 1 \longrightarrow y$ is a *fixed point* of α if $\alpha \circ p = p$. We are interested in morphisms $\alpha: y \longrightarrow y$ that do not have fixed points.

The main idea of Cantor's theorem is to determine if a certain function is representable in another function.

Definition 6.3. First a simple definition in **Set**. Let $f: S \times S \longrightarrow Y$ be a set function. The function f is a function of two variables. If s is an element in S, then $f(s,-): S \longrightarrow Y$ is a function of one variable. We say $g: S \longrightarrow Y$ is *representable* in f if there exists an s_0 in S such that $g(-) = f(s_0,-)$. In other words, for every x in S, we have that $g(x) = f(s_0, x)$. What does it mean for $g: S \longrightarrow Y$ to *not* be representable in f? It means for all s in S, $g(-) \neq f(s,-)$. In detail, for all $s \in S$, there is some x in S such that $g(x) \neq f(s, x)$.

Let us generalize representability to any category **A** with a terminal object 1 and binary products. We need the isomorphism $i: a \longrightarrow 1 \times a$. Let $f: a \times a \longrightarrow y$ and $g: a \longrightarrow y$ be morphisms in **A**. Then g is *representable* in f if there is a morphism $p: 1 \longrightarrow a$ such that

$$g = f \circ (p \times \mathrm{id}_a) \circ i: a \longrightarrow y \tag{6.7}$$

or

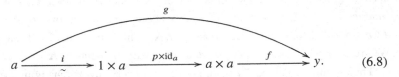

$$(6.8)$$

Then g is *not* representable in f if for all $p: 1 \longrightarrow a$ we have that $g \neq f \circ (p \times \mathrm{id}_a) \circ i$.

We can see this definition in terms of matrices. One can think of $f: a \times a \longrightarrow y$ as a matrix with entries in y. Then $f(s,-)$ is the sth row of the matrix. Given a $g: a \longrightarrow y$, we ask if it is some row of the matrix. It will be the case that g will be described so that it is different than every single row of f.

Now that we have all the ingredients, let us combine them and formulate the main theorem as in Lawvere (1969).

Theorem 6.4 (Cantor's Theorem). *Let A be a category with a terminal object and binary products. Let y be an object in A. If $\alpha: y \longrightarrow y$ is a morphism in A and α does not have a fixed point, then for every object a and for every $f: a \times a \longrightarrow y$, there exists a morphism $g: a \longrightarrow y$ such that g is not representable in f.*

Proof. The fact that there is a binary product means there is a morphism $\Delta: a \longrightarrow a \times a$. Let $\alpha: y \longrightarrow y$ not have a fixed point, then for any a, and for any $f: a \times a \longrightarrow y$ we can compose f with Δ and α to form g as in

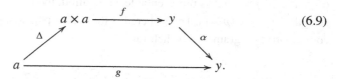

$$(6.9)$$

We claim that g is not representable in f. Assume (wrongly) that g is represented in f with $p: 1 \longrightarrow a$. This means that

$$g = f \circ (p \times \mathrm{id}_a) \circ i. \tag{6.10}$$

That is, for all $q: 1 \longrightarrow a$ we have

$$g \circ q = f \circ (p \times \mathrm{id}_a) \circ i \circ q. \tag{6.11}$$

For all q including $q = p$ that gives us

$$g \circ p = f \circ (p \times \mathrm{id}_a) \circ i \circ p \tag{6.12}$$

which shortens to

$$g \circ p = f \circ \Delta \circ p. \tag{6.13}$$

But by Diagram (6.9) the definition of g is

$$g \circ p = \alpha \circ f \circ \Delta \circ p. \tag{6.14}$$

Equations (6.13) and (6.14) imply that α has a fixed point:

$$(f \circ \Delta \circ p) = g \circ p = \alpha \circ (f \circ \Delta \circ p). \tag{6.15}$$

The fixed point for α is the map $(f \circ \Delta \circ p)$. Since α is assumed not to have any fixed point, the assumption that g is representable is wrong.   □

It is instructive to look at the proof in terms of matrices. The map g was constructed to be the changed diagonal of the matrix. If g is some row of the matrix, then we come to a contradiction. If g is the pth row of the matrix, then $g(-) = f(p,-)$. This is true for every element of the pth row of the matrix. The diagonal element is the (p,p)th element of the matrix. The map g was made to be different then $f(p,p)$, so they cannot be the same.

While most of the examples that we will see will be instances of Cantor's theorem, we will also have examples of instances of its contrapositive.

Theorem 6.5 (The Contrapositive of Cantor's Theorem). *Let* **A** *be a category with a terminal object and binary products. Let y be an object in* **A**. *If there exists an object a, and a morphism $f: a \times a \longrightarrow y$ such that every morphism $g: a \longrightarrow y$ is representable in f, then every $\alpha: y \longrightarrow y$ has a fixed point.*

Proof. Let **A**, y, a, and f be as described. Let $\alpha: y \longrightarrow y$ be any endomorphism of y. If $g = \alpha \circ f \circ \Delta$ is representable as required, then there is a $p: 1 \longrightarrow a$ such that $g = f \circ (p \times \mathrm{id}_a) \circ i$ (where i is the isomorphism $a \longrightarrow 1 \times a$.) This commutative diagram will be helpful:

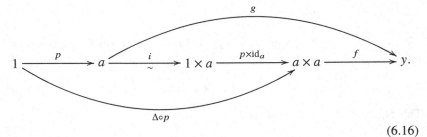

$$(6.16)$$

Since g is represented in f, and g is defined by $\alpha \circ f \circ \Delta$ we have both equalities

$$f \circ (p \times \mathrm{id}_a) \circ i = g = \alpha \circ f \circ \Delta. \qquad (6.17)$$

Precomposing all three maps by p gives us

$$f \circ (p \times \mathrm{id}_a) \circ i \circ p = g \circ p = \alpha \circ (f \circ \Delta \circ p). \qquad (6.18)$$

By Diagram (6.16), the left side shortens to

$$(f \circ \Delta \circ p) = g \circ p = \alpha \circ (f \circ \Delta \circ p). \qquad (6.19)$$

And here we see that $(f \circ \Delta \circ p)$ is a fixed point of α.   □

It is instructive to look at the proof in terms of matrices. If every function of one variable is some row of $f: a \times a \longrightarrow y$, then so is the g constructed to be the changed diagonal of the matrix. If g is the changed diagonal and the pth row of the matrix, then we have two values of $g(p) = f(p,p)$. These two values show that α has a fixed point.

Now let us apply these theorems.

Research Project 6.6. Cantor's theorem highlights the limitations that can be found in systems with self reference. It says that in such systems, there will be functions (g) that cannot be represented by such self-referential systems (f). One of the main ideas we saw in §5, was the notion of a reduction of decision problems. It would be nice to formulate the notion of a reduction of self-referential system. There are several levels of generality that are possible here. A simple example is as follows: given two self-referential systems $f \colon a \times a \longrightarrow y$ and $f' \colon a' \times a' \longrightarrow y$ a reduction is a map $h \colon a \longrightarrow a'$ such that the triangle

$$(6.20)$$

commutes. A more sophisticated notion of reduction might be between two self-referential systems in two different categories. Let $f \colon a \times a \longrightarrow y$ be a self-referential system in $\mathbf{A}$ and $f' \colon a' \times a' \longrightarrow y'$ be a self-referential system in $\mathbf{B}$. Then $H \colon \mathbf{A} \longrightarrow \mathbf{B}$ is a reduction of f to f' if $H(a) = a'$, $H(f) = f'$, $H(y) = y'$, and $H(\alpha \colon y \longrightarrow y) = \alpha \colon y' \longrightarrow y'$. There are probably many ways to weaken these requirements. Once we have such a reduction, how do unrepresentable functions in one system relate to unrepresentable functions in the other system? The fun will be finding examples of limitations that follow from such reductions.

6.2 Applications in Computability Theory

Example 6.7 (The Halting Problem Revisited). While we already saw in §4.1 that the Halting problem is not computable, it pays to see it again as an instance of Cantor's theorem.

First some reminders. We are discussing a Turing machine determining if a Turing machine with a certain input halts or goes into an infinite loop. To make things a little easier, we will only discuss Turing machines that take a single natural number as an input. As we saw in Diagram (3.50), we assign to every Turing machine a unique number that encodes it.

There is a function Halt: Nat × Nat ⟶ Bool defined as

$$\text{Halt}(x, y) = \begin{cases} 1: & \text{if Turing machine } y \text{ with input } x \text{ halts} \\ 0: & \text{if Turing machine } y \text{ with input } x \text{ goes on forever} \end{cases}$$

Theorem 6.8 (Turing's Unsolvability of the Halting Problem). *The Halt function is not a morphism in* **TotCompFunc**.

Proof. The f of Cantor's theorem will be the Halt function. Assume (wrongly) that Halt is in **TotCompFunc**, and hence in **CompFunc**. We can envision Halt as in Figure 6.15 where each row represents a Turing machine and the numbers on the top are the input numbers. The entries in the infinite matrix specify the values of Halt.

Halt	0	1	2	3	4	5	⋯
T_0:	0	1	0	0	0	1	⋯
T_1:	1	1	1	1	1	1	⋯
T_2:	0	1	0	0	0	0	⋯
T_3:	0	1	0	0	1	0	⋯
T_4:	0	0	1	1	1	1	⋯
T_5:	1	0	1	0	1	1	⋯
⋮	⋮			⋮		⋮	

Figure 6.15 The halting function.

Now consider the function Δ: Nat ⟶ Nat × Nat defined by $\Delta(x) = (x, x)$. This is clearly computable. The α of Cantor's theorem will be the "partial not" function ParNOT: Bool ⟶ Bool function which is defined as

$$\text{ParNOT}(x) = \begin{cases} 1: & \text{if } x = 0 \\ \uparrow: & \text{if } x = 1. \end{cases} \tag{6.21}$$

where $\uparrow$ means go into an infinite loop. ParNOT is not in **TotCompFunc**, but is in **CompFunc**. The g of Cantor's theorem will be the function Halt′ which will be given as follows

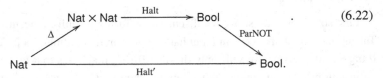

$$\tag{6.22}$$

Halt′ is in **CompFunc** since all of the morphisms it is composed of are in

CompFunc (Halt by assumption). Halt′ is defined as

$$\text{Halt}'(x) = \begin{cases} 1 : & \text{if } T_x \text{ on input } x \text{ does not halt, i.e., Halt}(x, x) = 0 \\ \uparrow : & \text{if } T_x \text{ on input } x \text{ does halt, i.e., Halt}(x, x) = 1 \end{cases}$$

Notice what Halt′ is doing. It is negating itself. If $\text{Halt}(x, x) = 1$ it is not halting and if $\text{Halt}(x, x) = 0$ then it halts. We can envision Halt′ as the function defined by the diagonal of Figure 6.16.

Halt	0	1	2	3	4	5	
T_0:	$\alpha(0) = 1$	1	0	0	0	1	$\cdots$
T_1:	1	$\alpha(1) = \uparrow$	1	1	1	1	$\cdots$
T_2:	0	1	$\alpha(0) = 1$	0	0	0	$\cdots$
T_3:	0	1	0	$\alpha(0) = 1$	1	0	$\cdots$
T_4:	0	0	1	1	$\alpha(1) = \uparrow$	1	$\cdots$
T_5:	1	0	1	0	1	$\alpha(1) = \uparrow$	$\cdots$
$\vdots$	$\vdots$			$\vdots$			$\vdots$

Figure 6.16 The changed diagonal of the purported halting function.

We claimed that Halt′ is a computable function and yet we saw that its output values are all different than every computable function's output values. Let us be more formal. If Halt′ was a computable function, then there would be some Turing machine, say the y_0 Turing machine, that would compute it. This means that the Halt′ function would halt on input x if and only if $\text{Halt}(y_0, x) = 1$. This should be true for all x. Let us ask Halt′ about itself by inputting y_0 into Halt′.

$$\text{Halt}'(y_0) = 1 \underset{\substack{\text{1 is output}}}{\Longleftrightarrow} \text{Halt}'(y_0) \text{ halts} \underset{\substack{\text{Halt}' \text{ number is } y_0}}{\Longleftrightarrow} \text{Halt}(y_0, y_0) = 1$$

$$\underset{\substack{\text{def. of Halt}'}}{\Longleftrightarrow} \text{Halt}'(y_0) = \uparrow . \tag{6.23}$$

This is clearly a contradiction and our assumption that Halt is computable was wrong. □

In terms of Figure 6.16, the proof says that Halt′, which is the changed diagonal, is not any row of the matrix and hence is not computable. The Halting function "diagonalizes out" of all total computable functions.

Example 6.9 (A computable function that is not primitive recursive). We saw in §3.3 that there is a subcategory **PRCompN** of **TotCompN** consisting of all primitive recursive functions. Advanced Topic 3.42 described the Ackermann function and stated (but did not prove) that it was an example of a morphism in **TotCompN** that is not in **PRCompN**. Here we will describe another morphism and prove that it is in **TotCompN**, and is not in **PRCompN**.

First some preliminaries about primitive recursive functions. To make things easier, we are going to only talk about computable functions that accept a single natural number and returns a single natural number, i.e., $\mathrm{Hom}_{\mathbf{CompN}}(\mathrm{Nat}, \mathrm{Nat})$. Such functions are called "unary." Just as we can encode every Turing machine as a unique natural number and every register machine as a unique natural number, so too, we can encode every unary primitive recursive function as a natural number. On page 39 we saw that the category of primitive recursive functions is generated by basic functions and the operations of composition and recursion. In a sense, a description of a primitive recursive function is like a program. It would say, take the input, do this basic function to the input, then do recursion with this and then compose with that function, etc. (One can view a description of a primitive recurisve function as a rooted tree where the leaves are the basic functions and the internal nodes are either an instance of composition or recursion. For more on this, see Yanofsky, 2011.) Regardless of the details, just as we can assign every computable function and every Turing machine a natural number, so too, we can assign to every unary primitive recursive function a natural number. For every natural number m, there is a unary primitive recursive function ϕ_m. If n is a natural number, then we shall write the value of ϕ_m with the input n as $\phi_m(n)$.

Theorem 6.10. *There is a total unary computable function that is not primitive recursive.*

Proof. This is an instance of Cantor's theorem, let $\mathbf{A} = \mathbf{TotCompN}$, $y = \mathrm{Nat}$, $\alpha \colon \mathrm{Nat} \longrightarrow \mathrm{Nat}$ defined as $\alpha(n) = n + 1$. This is the computable successor function and definitely does not have any fixed points. Now consider the function $f \colon \mathrm{Nat} \times \mathrm{Nat} \longrightarrow \mathrm{Nat}$ defined as $f(m, n) = \phi_m(n)$. The f function is in **TotCompN** because a Turing machine can decode the description m and then apply n to ϕ_m. Since all primitive recursive functions are total, f is a total function.

Define the function g as follows:

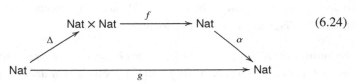

$$(6.24)$$

f	0	1	2	3	4	5	$\ldots$
0:	$\phi_0(0)+1$	$\phi_0(1)$	$\phi_0(2)$	$\phi_0(3)$	$\phi_0(4)$	$\phi_0(5)$	$\ldots$
1:	$\phi_1(0)$	$\phi_1(1)+1$	$\phi_1(2)$	$\phi_1(3)$	$\phi_1(4)$	$\phi_1(5)$	$\ldots$
2:	$\phi_2(0)$	$\phi_2(1)$	$\phi_2(2)+1$	$\phi_2(3)$	$\phi_2(4)$	$\phi_2(5)$	$\ldots$
3:	$\phi_3(0)$	$\phi_3(1)$	$\phi_3(2)$	$\phi_3(3)+1$	$\phi_3(4)$	$\phi_3(5)$	$\ldots$
4:	$\phi_4(0)$	$\phi_4(1)$	$\phi_4(2)$	$\phi_0(3)$	$\phi_4(4)+1$	$\phi_4(5)$	$\ldots$
5:	$\phi_5(0)$	$\phi_5(1)$	$\phi_5(2)$	$\phi_5(3)$	$\phi_5(4)$	$\phi_5(5)+1$	$\ldots$
$\vdots$	$\vdots$		$\vdots$		$\vdots$		

Figure 6.17 A listing of all descriptions of primitive recursive functions and their values with a changed diagonal. The diagonal computable function $g:$ Nat $\longrightarrow$ Nat is different from every description in the listing, and hence is not on the list.

which means that

$$g(n) = \alpha(f(n,n)) = \alpha(\phi_n(n)) = \phi_n(n) + 1. \qquad (6.25)$$

The function g is different from every description of a primitive recursive function. Formally, g is a computable function that is not primitive recursive. If g was primitive recursive, then it would have a description on the list, say n_0. That is, $g(m) = \phi_{n_0}(m)$ for all m. But this cannot be true because

$$g(n_0) = \phi_{n_0}(n_0) + 1 \neq \phi_{n_0}(n_0). \qquad (6.26)$$

So g "diagonalizes out" of the (descriptions of) primitive recursive functions. $\square$

It is instructive to understand this proof by understanding f and g as in Figure 6.17.

There is another related theorem that is proven the same way.

Theorem 6.11. *There is a total unary function that is not computable.*

Proof. The proof is almost exactly as the previous one. Rather than just considering descriptions of primitive recursive functions, consider descriptions of all total computable functions (using Turing machines, register machines, circuits, etc.) Then the diagonal is a total unary function of numbers that does not have a description as a computable function and hence is not computable. $\square$

Example 6.12. One of the deepest theorems in all of theoretical computer science is the *recursion theorem*.[8] The idea goes to the core of self reference and has many applications. While the theorem is usually painful to prove, with our categorical Cantor's theorem, it is rather simple.

First some preliminaries. We are only going to deal with descriptions of unary computable functions. As we have seen, there are many ways to describe unary computable functions. In order to fix our ideas, we will use register machines. We can encode register machine descriptions of unary computable functions as a natural number. That is, there is a computable isomorphism

$$\text{Hom}_{\textbf{RegMach}}(\text{Nat}, \text{Nat}) \longrightarrow d(\mathbb{N}) \tag{6.27}$$

where $d(\mathbb{N})$ is the discrete category of natural numbers. The m register machine will be denoted by R_m. The computable function that corresponds to the mth register machine will be denoted by φ_m. If we apply the number n to this function we get the $\varphi_m(n)$. If this is a number then we go one step further: we can consider $\varphi_m(n)$ to be the number of a register machine and look at that function, i.e., $\varphi_{\varphi_m(n)}$. This is a function describing a function, i.e., self reference at its finest. If $\varphi_m(n)$ is undefined, we consider $\varphi_{\varphi_m(n)}$. to be undefined.

Theorem 6.13 (The Recursion Theorem). *For every unary computable function* $h\colon \text{Nat} \longrightarrow \text{Nat}$*, there is a number n such that* $\varphi_{h(n)} = \varphi_n$.

Just to clarify, the theorem does not say that $h(n) = n$, i.e., that the descriptions are the same. Rather, the theorem says that the two (different) descriptions are describing the same function.

Proof. The proof is an instance of the contrapositive of Cantor's theorem. The category is $\mathbf{A} = \textbf{RegMach}$, $y = \text{Nat}^{\text{Nat}}$ (rather, the encoding of this), and $\alpha_h\colon \text{Nat}^{\text{Nat}} \longrightarrow \text{Nat}^{\text{Nat}}$ is defined by taking every register machine of a computable unary function R_x to $R_{h(x)}$. In Cantor's theorem a is Nat and f is the computable function $f\colon \text{Nat} \times \text{Nat} \longrightarrow \text{Nat}^{\text{Nat}}$ defined as $f(m, n) = \varphi_{\varphi_m(n)}$. This is a computable function because the encoding function is computable. Given m and n, a computer has to figure out R_m and then apply n. This gives number $\varphi_m(n)$. All that remains is to find the computable function that corresponds to that number. We have listed the outputs of f in Figure 6.18.

By composing f with Δ and α we can form $g\colon \text{Nat} \longrightarrow \text{Nat}^{\text{Nat}}$ defined as

$$g(m) = \varphi_{h(\varphi_m(m))} \tag{6.28}$$

and g is computable. There is some t such that $g(-) = f(t, -)$. (In terms of Figure 6.18, this means the diagonal is equal to row t.) One can actually

---

[8]In the literature, sometimes it is called "the second recursion theorem" and sometimes it is called "Rogers' version of Kleene's recursion theorem."

f	0	1	2	3	4	5	$\cdots$
0:	$\varphi_{h(\varphi_0(0))}$	$\varphi_{\varphi_0(1)}$	$\varphi_{\varphi_0(2)}$	$\varphi_{\varphi_0(3)}$	$\varphi_{\varphi_0(4)}$	$\varphi_{\varphi_0(5)}$	$\cdots$
1:	$\varphi_{\varphi_1(0)}$	$\varphi_{h(\varphi_1(1))}$	$\varphi_{\varphi_1(2)}$	$\varphi_{\varphi_1(3)}$	$\varphi_{\varphi_1(4)}$	$\varphi_{\varphi_1(5)}$	$\cdots$
2:	$\varphi_{\varphi_2(0)}$	$\varphi_{\varphi_2(1)}$	$\varphi_{h(\varphi_2(2))}$	$\varphi_{\varphi_2(3)}$	$\varphi_{\varphi_2(4)}$	$\varphi_{\varphi_2(5)}$	$\cdots$
3:	$\varphi_{\varphi_3(0)}$	$\varphi_{\varphi_3(1)}$	$\varphi_{\varphi_3(2)}$	$\varphi_{h(\varphi_3(3))}$	$\varphi_{\varphi_3(4)}$	$\varphi_{\varphi_3(5)}$	$\cdots$
4:	$\varphi_{\varphi_4(0)}$	$\varphi_{\varphi_4(1)}$	$\varphi_{\varphi_4(2)}$	$\varphi_{\varphi_0(3)}$	$\varphi_{h(\varphi_4(4))}$	$\varphi_{\varphi_4(5)}$	$\cdots$
5:	$\varphi_{\varphi_5(0)}$	$\varphi_{\varphi_5(1)}$	$\varphi_{\varphi_5(2)}$	$\varphi_{\varphi_5(3)}$	$\varphi_{\varphi_5(4)}$	$\varphi_{h(\varphi_5(5))}$	$\cdots$
$\vdots$	$\vdots$			$\vdots$		$\vdots$	

Figure 6.18 A listing of all unary computable functions defined by $\varphi_m(n)$. The diagonal is changed by a computable function h. At least one row will be the same as the diagonal. That will provide a fixed point.

determine which t by looking at the SMN theorem.) This is true for all input including t, hence

$$g(t) = f(t,t) = \varphi_{\varphi_t(t)}. \tag{6.29}$$

(In terms of Figure 6.18, this means that the (t,t) position has two different definitions that are equivalent.) Plugging t into Equation (6.28) gives us

$$g(t) = \varphi_{h(\varphi_t(t))}. \tag{6.30}$$

Setting $n = \varphi_t(t)$ and combining Equations (6.29) and (6.30) results in $\varphi_{h(n)} = \varphi_n$. $\qquad\Box$

The recursion theorem is usually used in conjunction with the SMN theorem as follows.

Theorem 6.14. *Let* $h \colon Nat \times Nat \longrightarrow Nat$ *be any computable function. Then there exists a natural number e such that* $\varphi_e(y) = h(e,y)$.

Proof. Let h be given. By the SMN theorem there is a computable function $s(x)$ such that $\varphi_{s(x)} = h(x,y)$. Apply the recursion theorem to s and let e be the fixed point. So we have $\varphi_e(y) = \varphi_{s(e)}(y) = h(e,y)$. $\qquad\Box$

There are many applications of the recursion theorem:

- One can prove a version of Turing's unsolvability of the Halting problem with the recursion theorem.
- One can prove Rice's theorem with the recursion theorem.
- There is a natural number e such that $\varphi_e(x) = x^e$. (Simply set h of Theorem 6.14 to be $h(y,x) = x^y$.)

- There is a program that can output its own description. This means there is a number, e, of a program that only outputs e: $\varphi_e(x) = e$. (Simply set h of Theorem 6.14 to be $h(y, x) = y$.) This is related to John von Neumann's *self-replicating machine*.
- In algebraic topology, there is an interesting result that goes back to Henri Poincaré called "the hairy ball theorem." The theorem essentially says that for any way one combs the hair on a ball, there will always be at least one place where the hair is straight up. (A physical application of this theorem is the realization that at every instance, there is at least one place at the surface of the Earth where the wind is *not* blowing.) Note, there is no problem combing the hair on a flat plain or on a doughnut. The literature has an application of the recursion theorem that is essentially a computer science version of the hairy ball theorem. Imagine wanting to change an infinite number of programs in a computable way. We would do this with computable function $h\colon \text{Nat} \longrightarrow \text{Nat}$. We change the mth program by changing it into the $h(m)$th program. That is, $\varphi_m \longmapsto \varphi_{h(m)}$. The recursion theorem says that this cannot be done. There will be some n such that $\varphi_n = \varphi_{h(n)}$. You can systematically change an infinite number of programs but you cannot systematically change an infinite number of computable functions.

It should be noted that Gödel's incompleteness theorem can also be seen as an instance of the contrapositive of Cantor's theorem. However, since we have already seen this important theorem as a consequence of the unsolvability of the Halting problem in §4.2, we leave it out. See Yanofsky (2003) for all the gory details.

Research Project 6.15. There are actually several different variations of the recursion theorem. It would be nice to see if any of the other ones can also be written as an instance of the contrapositive of Cantor's theorem.

6.3 Applications in Complexity Theory

Example 6.16 (Space Hierarchy). The more resources you have, the more problems you will be able to solve. We know from the end of §5.1 that if $o(h) \subsetneq O(h)$ then **DTIME**$(o(h)) \subseteq$ **DTIME**$(O(h))$ and **DSPACE**$(o(h)) \subseteq$ **DSPACE**$(O(h))$, but here we show that for certain h these inclusions of categories are proper inclusions.

Definition 6.17. A computable function $h\colon \mathbb{N} \longrightarrow \mathbb{R}^*$ is called *space-constructable* if there is a Turing machine that takes n to $\lceil h(n) \rceil$ in binary and does not use more than $\lceil h(n) \rceil$ space. We say h is *time-constructable* if there is a Turing machine that takes n to $\lceil h(n) \rceil$ in binary and does not use more than $\lceil h(n) \rceil$ time clicks. Most typical computable functions satisfy these requirements.

Theorem 6.18 (Space Hierarchy Theorem). *Let h be a space constructable function, then* **DSPACE**$(o(h))$ *is a proper subcategory of* **DSPACE**$(O(h))$.

Proof. We will describe a decision problem g: String $\longrightarrow$ Bool that is in **DSPACE**$(O(h))$ but is not in **DSPACE**$(o(h))$.

The proof is an instance of Cantor's theorem. The category is **TotCompFunc**. The object y = Bool and α: Bool $\longrightarrow$ Bool is the NOT function which has no fixed point. The a object is String. Rather than defining f: String $\times$ String $\longrightarrow$ Bool for strings M and w, we shall define $\alpha \circ f$ since it will be more understandable for the rest of the proof.

$$
\alpha(f(M,w)) = \begin{cases} \text{accept:} & \begin{array}{l} \text{if } M \text{ is a description of a Turing machine that} \\ \text{rejects } w \text{ within } o(h(|w|)) \text{ space.} \end{array} \\ \text{reject:} & \text{otherwise.} \end{cases}
$$

The technical part of this proof is showing that $\alpha \circ f$ is computable within $O(h(|w|))$ and hence in **DSPACE**$(O(h))$. This is true because of the following facts.

- A Turing machine can easily determine if an input is a valid description of a Turing machine. This does not demand a lot of time or space.
- A Turing machine can easily make sure that while simulating a Turing machine on an input, the simulation does not use more than $h(|w|)$ space. This is done by using the space-constructability of h and marking off that much space on the work tape. At any point while simulating the Turing machine, if the simulation attempts to go beyond the marking, then the simulation is using too much space.
- A Turing machine can easily make sure that a space bounded simulation does not go into an infinite loop. If the simulated Turing machine has k elements in the alphabet, then if the simulation goes beyond $k^{h(n)}$ steps, the simulation is repeating a configuration and is in an infinite loop. Hence the Turing machine has to keep a running counter of how many steps it has performed, and when the counter goes beyond $k^{h(n)}$, it knows the simulation is in infinite loop and therefore should halt. If the counter is on another tape, this will demand $k^{h(n)}$ time and $h(n)$ space for the counter.

Let $g = \alpha \circ f \circ \Delta$. Then g : String $\longrightarrow$ Bool is defined as

$$g(M) = \begin{cases} \text{accept:} & \begin{aligned}&\text{if } M \text{ is a description of a Turing machine that } \underline{\text{rejects}} \\ & \qquad\qquad M \text{ within } o(h(|M|)) \text{ space.}\end{aligned} \\ \text{reject:} & \text{otherwise.} \end{cases}$$

Then g is made out of the composition of three morphisms (f, Δ, and NOT) and their composition uses $O(h(n))$ space but the main point is that $M(M)$ rejects within $o(h(|M|))$ space if and only if $g(M)$ accepts. We conclude that there is no Turing machine that accepts what g accepts in $o(h(n))$ space and therefore the language that g accepts is not in $o(h(n))$ space.   □

There is also a nondeterministic space hierarchy theorem which we will not go into.

Example 6.19 (Time Hierarchy). This is similar to the previous theorem with one added complexity.

Theorem 6.20 (Time Hierarchy Theorem). *For time constructable h,* **DTIME**$(o(h(n)/(\log h(n))))$ *is a proper subcategory of* **DTIME**$(O(h))$

Proof. The proof starts off almost exactly like the last one. We will describe a decision problem g : String $\longrightarrow$ Bool that is in **DTIME**$(O(h))$ but is not in **DTIME**$(o(h(n)/(\log h(n))))$.

The proof is an instance of Cantor's theorem. The category is **TotCompFunc**. The object y = Bool and α = NOT. The a object is String and f : String $\times$ String $\longrightarrow$ Bool is defined for strings M and w. For ease of proof, we will define $\alpha \circ f$ as follows:

$$\alpha(f(M, w)) = \begin{cases} \text{accept:} & \begin{aligned}&\text{if } M \text{ is a description of a TM that } \underline{\text{rejects }} w \\ & \qquad\qquad \text{within } o(h(n)/(\log h(n))) \text{ time.}\end{aligned} \\ \text{reject:} & \text{otherwise.} \end{cases}$$

The technical part of this proof is showing that f is computable within $O(h(|w|))$ time. This is true because of the following facts.

- A Turing machine can easily determine if an input is a valid description of a Turing machine. This does not demand a lot of time or space.
- A Turing machine can easily make sure that a simulation stays within a time bound. Use the time constructability to find $h(|w|)/(\log h(|w|))$ and put it into a counter on an unused part of a work tape. At every time click, the counter is decremented. If the counter ever hits 0, then the simulation has

gone beyond the bound. The division by $\log h(|w|)$ comes from the fact that the Turing machine has to go back and forth from the time counter.[9]

Let $g = \alpha \circ f \circ \Delta$. g: String $\longrightarrow$ Bool which is defined as

$$g(M) = \begin{cases} \text{accept:} & \begin{array}{l}\text{if } M \text{ is a description of a TM that } \overline{\text{rejects}} \ M \\ \text{within } o(h(|M|)/(\log h(|m|))) \text{ time.}\end{array} \\ \text{reject:} & \text{otherwise.} \end{cases}$$

Then g is made out of the composition of three morphisms (f, Δ, and NOT). Their composition is in $O(h(n))$ time but the main point is that $M(M)$ rejects within $o(h(|M|/(\log(h()))$ time if and only if $g(M)$ accepts. We conclude that there is no Turing machine that accepts what g accepts within the proper amount of time. $\qquad\square$

There is also a nondeterministic time hierarchy theorem which we will not go into.

Example 6.21. In §4.3 we discussed the notion of an oracle Turing machine and used it to help us classify unsolvable problems. We can use oracles in complexity theory also. Let f be a decision problem f: Seq $\longrightarrow$ Bool, then we talk about oracle Turing machine $M^?$ with the decision problem f as an oracle and write this as M^f. We usually think of f as some more complicated function and the Turing machine $M^?$ can freely ask questions of f. We extend this notation even further. If **A** is a complexity class of decision problems then $\mathbf{A}^f$ is the complexity class of all the decision problems that can be solved by the Turing machines that represent functions in **A** which can freely use the information of f.

The big open question in complexity theory is whether or not **P** is equal to **NP**. The following deals with the analogous question concerning oracle computations.

Theorem 6.22 (Baker–Gill–Solovay Theorem). *There exists oracles A: Seq $\longrightarrow$ Bool and B: Seq′ $\longrightarrow$ Bool such that $\mathbf{P}^A = \mathbf{NP}^A$ and $\mathbf{P}^B \neq \mathbf{NP}^B$.*

Proof. The first statement is rather easy. For any decision problem A, since **P** is a subcategory of **NP**, it is obvious that $\mathbf{P}^A$ is a subcategory of $\mathbf{NP}^A$. We are left with the task to find an A such that $\mathbf{NP}^A$ is a subcategory of $\mathbf{P}^A$. Let A be any PSPACE-complete decision problem. Then we have

$$\mathbf{NP}^A \overset{1}{\hookrightarrow} \mathbf{NPSPACE} \overset{2}{\underset{=}{\hookrightarrow}} \mathbf{PSPACE} \overset{3}{\hookrightarrow} \mathbf{P}^A. \qquad (6.31)$$

[9]Since we have many work tapes, we can put the time counter on a totally different tape. In this case there is no traveling back and forth from the time counter. One hand of the Turing machine will always be on the time counter. The theorem then becomes simpler: For time constructable h, **DTIME**$(o(h(n)))$ is a proper subcategory of **DTIME**$(O(h))$.

Inclusion 1 comes from the fact that every polynomial time nondeterministic computation calls its oracle at most a polynomial amount of times. Rather than calling the oracle each time, we can use polynomial space to simulate the execution of A. This will still be a polynomial amount of space and hence will be in **NPSPACE**. Inclusion 2 is simply the content of Savitch's theorem from §5.3. Inclusion 3 comes from the fact that A is PSPACE-complete and for any decision problem $g\colon \text{Seq}' \longrightarrow \text{Bool}$ in **PSPACE**, by PSPACE-completeness there is a deterministic polynomial map $h_g\colon \text{Seq}' \longrightarrow \text{Seq}$ such that $g = A \circ h_g$ as in

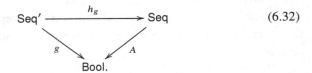

$$\text{(6.32)}$$

Rather than solving g, we can run a deterministic program that executes h_g and then calls the oracle A once. This program is in $\mathbf{P}^A$.

The next statement is much harder and uses a diagonal argument. We will use the diagonal argument to construct an oracle B and a language L_B that is in $\mathbf{NP}^B$ but "diagonalizes out" of $\mathbf{P}^B$. To make things easier, both B and L_B will be computable functions of the form $\text{Bool}^* \longrightarrow \text{Bool}$. That is, they both describe languages in $\{0,1\}$. The language L_B will be intimately related to B as follows

$$L_B = \{1^n : n = |b| \text{ for some } b \in B\}. \qquad (6.33)$$

That is, L_B equals the lengths of the elements in B.

Technical Point 6.23. For those who know and love the language of Kan extensions, L_B is the Kan extension of B along the length functor $|-|$ as in

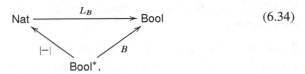

$$\text{(6.34)}$$

Given an input 1^n, in order to determine if 1^n is in L_B, a nondeterministic Turing machine must guess one of the 2^n elements of $\{0,1\}^n$ and use his oracle to check if any of them are in B. If at least one of them is in B, then the machine should accept 1^n. Otherwise, reject. This shows that L_B is in $\mathbf{NP}^B$.

Now we have to show that L_B is not in $\mathbf{P}^B$. We do this with a diagonal argument. First there is an encoding of polynomial time oracle Turing machines $M_0^?, M_1^?, M_2^?, \ldots$. Each of these Turing machines has an associated polynomial that describes its running time: $p_0, p_1, p_2, \ldots$. We will be diagonalizing out of these $M_i^?$. Our goal is to create an oracle set (characteristic function) B that is built up in stages:

$$\emptyset = B_0 \subseteq B_1 \subseteq B_2 \subseteq \cdots \subseteq B. \tag{6.35}$$

We define $f : \mathsf{Nat} \times \mathsf{Nat} \longrightarrow \mathsf{Bool}$ as follows:

$$f(m,n) = \begin{cases} 1: & \text{if oracle Turing machine } M_m^{B_{n-1}} \text{ accepts } 1^t \\ 0: & \text{if oracle Turing machine } M_m^{B_{n-1}} \text{ rejects } 1^t. \end{cases} \tag{6.36}$$

where t is defined as the first number that is larger than any element in B_{n-1} and $2^t > p_m(t)$. We are either going to add an element of size t to B or not. Because $2^t = |\{0,1\}^t|$ is larger than $p_m(t)$, there are more possible words to add to B than there are words that $M_m^?$ can ask the oracle. Define α by the now-familiar NOT: $\mathsf{Bool} \longrightarrow \mathsf{Bool}$. Define $\Delta : \mathsf{Nat} \longrightarrow \mathsf{Nat} \times \mathsf{Nat}$ as $\Delta(n) = (n,n)$. By composition $g : \mathsf{Nat} \longrightarrow \mathsf{Bool}$ is defined as

$$g(n) = \begin{cases} 1: & \text{if oracle Turing machine } M_n^{B_{n-1}} \text{ rejects } 1^t \\ 0: & \text{if oracle Turing machine } M_n^{B_{n-1}} \text{ accepts } 1^t. \end{cases} \tag{6.37}$$

We can view f and the diagonal g as in Figure 6.19

f	B_0	B_1	B_2	B_3	B_4	B_5	$\cdots$
$M_0^?$:	$\alpha(0) = 1$	1	0	0	0	1	$\cdots$
$M_1^?$:	1	$\alpha(1) = 0$	1	1	1	1	$\cdots$
$M_2^?$:	0	1	$\alpha(0) = 1$	0	0	0	$\cdots$
$M_3^?$:	0	1	0	$\alpha(0) = 1$	1	0	$\cdots$
$M_4^?$:	0	0	1	1	$\alpha(1) = 0$	1	$\cdots$
$M_5^?$:	1	0	1	0	1	$\alpha(1) = 0$	$\cdots$
$\vdots$	$\vdots$			$\vdots$			$\vdots$

Figure 6.19 The changed diagonal g of the f function.

The main point is to use g to define each stage of B. Let x be a word such that $|x| = t$ and x was not queried by $M_n^{B_{n-1}}$ on input 1^t.

$$B_{n+1} = \begin{cases} B_n : & \text{if } g(n) = 0 \text{ i.e., } M_n^{B_{n-1}} \text{ accepts } 1^t, \text{ i.e., } 1^t \notin L_B \\ B_n \cup \{x\} : & \text{if } g(n) = 1 \text{ i.e., } M_n^{B_{n-1}} \text{ rejects } 1^n, \text{ i.e., } 1^t \in L_B, \end{cases}$$

where g is the characteristic function of the elements of L_B and Cantor's theorem tells us that g is not on our list of deterministic polynomial oracle Turing machines. $\square$

Advanced Topic 6.24. Hartmanis and Hopcroft (1976) showed that there exists an oracle C such that the question of $\mathbf{P}^C = \mathbf{NP}^C$ is independent of the Zermelo–Frankel axioms of set theory. This is proven by essentially "diagonalizing out" of all proofs using the axioms of Zermelo–Frankel set theory.

It should be noted that Ladner's theorem mentioned in Advanced Topic 5.23 is also proven using a diagonal argument. (See Section 3.3 in Arora and Barak, 2009.)

With all this "diagonalizing out" one might think that the solution to the $\mathbf{P} = \mathbf{NP}$ question can easily be solved with a diagonal argument. It seems that all we have to do in order to show that $\mathbf{P} \neq \mathbf{NP}$ is find a decision problem $f : \mathsf{Seq} \longrightarrow \mathsf{Bool}$ that is solvable in $\mathbf{NP}$ but "diagonalizes out" of $\mathbf{P}$. While it seems simple, the Baker–Gill–Solovay theorem shows us that this is probably not true. If there were such a proof, then it could easily be upgraded to a proof about oracle Turing machines for any set X. This upgraded proof will show that for any oracle X we have $\mathbf{P}^X \neq \mathbf{NP}^X$. However, we now know that for certain oracles the two classes are equal and for other oracles the two classes are not equal. This shows a deep limitation of the diagonal method. (Not everyone is convinced by this argument.)

Research Project 6.25. Here are a few small potential research projects:

- Formulate the nondeterministic space and nondeterministic time hierarchy theorems as instances of Cantor's theorem.
- We saw that given complexity classes $\mathbf{A}$ and $\mathbf{B}$, one can form the complexity class $\mathbf{A}^{\mathbf{B}}$ which consists of all functions that can be computed with machines in $\mathbf{A}$ that use oracles in $\mathbf{B}$. It would be nice to formulate this mapping as a type of bifunctor from pairs of complexity classes to complexity classes. What properties does this functor have?
- Understand the result from Advanced Topic 6.24 and formulate it as an instance of Cantor's theorem.

Further Reading

- Turing's Halting problem: Section 5.3 of Mendelson (1997), Section 4.2 of Sipser (2006), Chapter V of Manin (2010), Section 4.1 of Boolos et al. (2007), and Section 6.2 of Yanofsky (2013).
- For computable functions that are not primitive recursive see Section 4.9 of Davis et al. (1994b), Section 10.3 of Cutland (1980), and Appendix A of Eilenberg and Elgot (1970).
- The recursion theorem and its consequences can be found Chapter 11 of Cutland (1980), Section 4.8 of Davis et al. (1994b), and Chapter 11 of Rogers

(1987).

- The space and time hierarchy theorems can be found in Section 9.1 of Sipser (2006), Section 29.6 of Rich (2008), Chapter 3 of Arora and Barak (2009), and Section 12.5 of Hopcroft and Ullman (1979).
- The Baker–Gill–Solovay Theorem can be found in Section 9.2 of Sipser (2006), Section 13.7 of Hopcroft and Ullman (1979), and Section 14.3 of Papadimitriou (1994).

The idea that category theory can describe all these diverse self-referential phenomena was brilliantly formulated in Lawvere (1969). These ideas are discussed in Session 29 of Lawvere and Schanuel (2009) and in Section 7.3 of Lawvere and Rosebrugh (2003). Yanofsky (2003) has the goal of making these ideas understandable for people who are novices to category theory. The paper contains nineteen instances of Cantor's theorem and its contrapositive from all over logic, mathematics, and theoretical computer science.

A larger discussion of all these self-referential paradoxes for the general reader can be found in Yanofsky (2013): see Chapter 10, and in particular, page 344 for a unified listing of many self-referential paradoxes.

7 Conclusion

7.1 Looking Back

We have come to the end of our journey. I now want to summarize our work by examining all the different types of categorical structures we have repeatedly used throughout the text. We also highlight what was gained by looking at theoretical computer science from the category-theoretic point of view.

- Rather than just looking at the set of all functions, we described the symmetric monoidal category of functions and various symmetric monoidal subcategories that are of interest. The symmetric monoidal categorical structure highlights the sequential and parallel composition of the functions §3.1.)
- Rather than looking at a set of models of computation, we described the symmetric monoidal bicategory which have models of computation and various symmetric monoidal sub-bicategories that are of interest. This highlights the sequential and parallel composition of the models of computation. Such structures were used to hold

  - Turing machines (§3.2),
  - Register machines (§3.3),
  - Circuits (§3.4), and
  - Logical formulas (§3.5).

- All the functors relating functions, models of computation, and logical formulas are symmetric monoidal functors and hence respect sequential and parallel processing (§3).
- Every function that is computable with one model of computation has an equivalent other model of computation. (This is the content of Figure 3.8.)
- Rather than considering the set of isolated decision problems, we worked with the category of decision problems and their reductions. They are comma categories. In complexity theory, these categories form complexity classes (§§5.2 and 5.3).
- Complete problems for a complexity class are weak terminal objects in the category of decision problems. They form a subcategory of the decision problems (Definition 5.19 and §§5.2 and 5.3).
- There are a few major results (e.g., Halting is undecidable, SAT is NP-Complete, the recursion theorem, etc.) and many corollaries are derived from those results. This follows from the fact that decision problems are a comma category where it is easy to express reductions.
- We described a single categorical theorem about self-referential systems and there were many instances of this theorem in computability and complexity theory (§6).
- We gave a simple categorical definition of the category of finite automata (§1 of the supplement).
- We gave a simple categorical definition of a cryptographic protocol and showed that every major crypotographic protocol is an instance of this definition (§2 of the supplement).
- We showed that looking at programs, algorithms, and functions as categories and the functors between them clarify the relationship between these domains (§4 of the supplement).
- Adjoint functors *did not* play a major role in the tale that we've told. In computer science, there are many equivalent ways of making constructions. This is not conducive to universal properties.
- We, however, did use Kan extensions as ways of finding complicated minimization and maximazation functors (Technical Points 5.3, 5.4, 6.23, and in §3 of the supplement)
- In §3.5, we defined a functor L_t from the symmetric monoidal bicategory of total Turing machines to the symmetric monoidal bicategory of families of sequences of logical formula. This functor was used to describe the workings of a Turing machine with logical formulas. The functor L_t and extensions of the functor were used in the proofs of the following theorems that relate computation and logic:

– Gödel's Incompleteness Theorem. (Theorem 4.15).
– The unsolvability of the Entscheidungsproblem (Theorem 4.16).
– The Cook–Levin Theorem. (Theorem 5.20).

• Many times we saw that a subcategory was "dense" inside a category. That means there is an inclusion Inc of a subcategory **A** into a category **B** such that every b in **B** is similar to some a in **A**. Such density notions came in two types.

– Inc is full, faithful and inherently surjective. (Definition 3.14). For example,

  * Every computable function has an equivalent computable function between powers of String (Theorem 3.15).
  * Every computable function has an equivalent computable function between powers of Nat (Theorem 3.24).
  * Every computable function has an equivalent computable function between powers of Bool* (Theorem 3.45).

– Another way to talk about density can be described with a diagram of the form

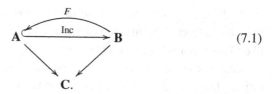

$$(7.1)$$

where both triangles commute and $F \circ \text{Inc} = \text{Id}_{\mathbf{A}}$ but, in general $\text{Inc} \circ F \neq \text{Id}_{\mathbf{B}}$. What this means is that for every b in **B**, $F(b)$ is not the same as an a in **A**, but is the same in relation to the functors to **C**. For example,

  * Every Turing machine is equivalent to a Turing machine in **Turing**$(1, 1)$ (Diagram (3.21)).
  * Every logic circuit has an equivalent NAND logic circuit (Diagram (3.44)).
  * Every nondeterministic Turing machine has an equivalent deterministic Turing machine (Diagram (5.28)).
  * Every **NPSPACE** computation has an equivalent **PSPACE** computation (Theorem 5.27, i.e., Savitch's theorem).
  * Every **NP**^{A} computation is equivalent to a **NPSPACE** computation where A is PSPACE-complete (Diagram (6.31)).
  * Every **PSPACE** computation is equivalent to a **P**^{A} computation where A is PSPACE-complete (Diagram (6.31)).
  * Every nondetermistic finite automaton has an equivalent deterministic

finite automaton[10] (§1 of the supplement).

- While our aim was to be as categorical as possible, we found that twice we
  had to go outside the bounds of categories:

  – Encodings of models of computation or of computable functions are
    not functorial. They neither respect sequential composition nor parallel
    composition (§3.6).
  – Complexity measures of models of computation or of computable functions
    are not functorial. They neither respect sequential composition nor parallel
    composition (Technical Point 5.5).

Of course, the devil is in the details for each case. However, the uniform
categorical description is helpful.

The positive effects of the "categorical imperative" (i.e., the absolute, un-
conditional urge to state truths in the language of category theory) is plainly
evident in this list.

7.2 Moving Forward

We have only scratched the surface. Theoretical computer science is an
immense subject. Now that you have the basics of this beautiful topic, we would
like to point out ways for you to go on.

Some textbooks in theoretical computer science are Sipser (2006), Rich
(2008), Davis et al. (1994b), Lewis and Papadimitriou (1997), Boolos et al.
(2007), Cutland (1980), and Rogers (1987).

For a popular, non-technical introduction to many of these ideas, see Chapters
5 and 6 of Yanofsky (2013) and Harel (2000). Aaronson (2013) is a delightful
popular text that takes some of these ideas further and relates them to parts
of physics. Harrow (1979) is a beautiful article on the relationship between
theoretical and applied computer science.

None of the above sources mention any category theory. Our presentation
is novel in that these topics have not been presented before in a uniform way
using categories. There is still much to learn about the relationship between
category theory and computer science. A good place to start looking are in some
textbooks and papers such as Barr and Wells (1999), Walters (1991), Pierce
(1991), Asperti and Longo (1991), Scott (2000), Spivak (2014), Poigné (1992),
Ehrig et al. (1974), and Fiadeiro (2004).

---

[10]It should be pointed out that this relationship between nondeterminism and determinism is
not universal. Here are two examples where it fails. (i) Not every nondeterministic pushdown
automaton is equivalent to a deterministic pushdown automaton. (ii) Also, if $\mathbf{P} \neq \mathbf{NP}$ then not every
nondeterministic polynomial algorithm has an equivalent deterministic polynomial algorithm.

This text uses category theory as a bookkeeping tool to store and compare all the various parts of theoretical computer science. There is, however, a branch of research that uses category theory in a deeper way. They describe properties of categories that would be able to deal with computations. Perhaps the first work in this direction was done by one of the founders of category theory, Sammy Eilenberg. Towards the end of his career, in 1970, he and Calvin C. Elgot published a small book titled *Recursiveness* (Eilenberg and Elgot, 1970). In 1974 and 1976 he published a giant two-volume work on formal language theory titled *Automata, Languages, and Machines* (Eilenberg, 1974, 1976). Giuseppe Longo and Eugenio Moggi also had several papers in this direction (see Longo and Moggi, 1984b,a, 1990). Seely (1987) is an interesting paper about looking at a computation from a 2-categorical point of view.

Alex Heller (who was my thesis advisor and a dear friend) and Robert DiPaola (a teacher and a dear friend) published Di Paola and Heller (1987). This work was followed by Heller (1990) and Lengyel (2004). There are various types of similar categories with names like "P-Categories," "Restriction categories," " Recursion categories," and "Turing Categories." See Cockett and Hofstra (2008) and Hofstra and Scott (2021) for clear histories of the development of these ideas. Dusko Pavlovic develops the notion of a computation in a monoidal category in a series of papers that begin with Pavlovic (2013). There is also a development of such ideas for complexity theory in "Otto's thesis" (Otto, 1995) and in Diaz-Boïls (2017).

Manin (2000, 2010) put all computations into one category called a "computational universe." He was also able to incorporate quantum computing into his categories.

Another connection between theoretical computer science and category theory is implementing categorical structures on computers. The first place to look for this is in Rydeheard and Burstall (1988).

It is worth mentioning yet another connection between theoretical computer science and category theory: Yanofsky (2015) showed that there are constructions in category theory that can perform the workings of a Turing machine. Since limits and colimits are infinitary operations, it is possible for categories to "solve" the Halting problem (but this solution cannot be implemented on a finite computer).

References

Aaronson, Scott. 2013. *Quantum Computing Since Democritus*. Cambridge University Press.

Agrawal, Manindra, Kayal, Neeraj, and Saxena, Nitin. 2002. PRIMES is in P. *Ann. of Math*, **2**, 781–793.

Arora, Sanjeev, and Barak, Boaz. 2009. *Computational Complexity: A Modern Approach*. Cambridge University Press.

Asperti, Andrea, and Longo, Giuseppe. 1991. *Categories, Types, and Structures: An Introduction to Category Theory for the Working Computer Scientist*. MIT Press.

Baez, J., and Stay, M. 2011. Physics, topology, logic and computation: a Rosetta Stone. Pages 95–172 of: *New Structures for Physics*, Bob Coecke (ed). Lecture Notes in Physics, vol. 813. Springer.

Barr, Michael, and Wells, Charles. 1985. *Toposes, Triples and Theories*. Grundlehren der Mathematischen Wissenschaften, vol. 278. Springer.

Barr, Michael, and Wells, Charles. 1999. *Category Theory for Computing Science*, Third edition. Les Publications CRM, Montreal.

Bénabou, Jean. 1967. Introduction to bicategories. Pages 1–77 of: *Reports of the Midwest Category Seminar*. Lecture Notes in Mathematics, vol. 47. Springer.

Boolos, G.S., Burgess, J.P., and Jeffrey, R.C. 2007. *Computability and Logic*, Fifth edition. Cambridge University Press.

Cockett, J.R.B., and Hofstra, P.J.W. 2008. Introduction to Turing categories. *Ann. Pure Appl. Logic*, **156**(2–3), 183–209.

Coecke, Bob, and Kissinger, Aleks. 2017. *Picturing Quantum Processes: A First Course in Quantum Theory and Diagrammatic Reasoning*. Cambridge University Press.

Cormen, Thomas H., Leiserson, Charles E., Rivest, Ronald L., and Stein, Clifford. 2009. *Introduction to Algorithms*, Third edition. MIT Press.

Cutland, Nigel. 1980. *Computability: An Introduction to Recursive Function Theory*. Cambridge University Press.

Davis, M., and Hersh, R. 1973. Hilbert's 10th problem. *Scientific American*, **229** (Nov.), 84–91.

Davis, Martin, Segal, Ron, and Weyuker, Elaine. 1994a. *Computability, Complexity and Languages*. Academic Press.

Davis, Martin D., Sigal, Ron, and Weyuker, Elaine J. 1994b. *Computability, Complexity, and Languages*, Second edition. Academic Press.

Dean, Walter. 2016. Computational Complexity Theory. In: *The Stanford Encyclopedia of Philosophy*, winter 2016 edition, Zalta, Edward N. (ed). Metaphysics Research Lab, Stanford University.

Di Paola, Robert A., and Heller, Alex. 1987. Dominical categories: recursion theory without elements. *J. Symbolic Logic*, **52**(3), 594–635.

Diaz-Boïls, Joaquim. 2017. Categorical comprehensions and recursion. *Journal of Logic and Computation*, **27**, 1607–1641.

Eilenberg, Samuel. 1974. *Automata, Languages, and Machines. Vol. A.* Academic Press.

Eilenberg, Samuel. 1976. *Automata, Languages, and Machines. Vol. B.* Academic Press.

Eilenberg, Samuel, and Elgot, Calvin. 1970. *Recursiveness.* Academic Press.

Fiadeiro, José Luiz. 2004. *Categories for Software Engineering.* Springer.

Garey, M.R., and Johnson, D.S. 1979. *Computers and Intractability: A Guide to the Theory of NP-Completeness.* W.H. Freeman & Co.

Gurski, Nick. 2011. Loop spaces, and coherence for monoidal and braided monoidal bicategories. Available at arXiv:1102.0981.

Ehrig, H., Kiermeier, K.-D., Kreowski, H.-J., and Kühnel, W. 1974. *Universal Theory of Automata. A Categorical Approach.* Teubner.

Harel, David. 2000. *Computers Ltd.: What They Really Can't Do.* Oxford University Press, Inc.

Harrow, Keith. 1979. Theoretical and applied computer science: antagonism or symbiosis? *Amer. Math. Monthly,* **86**(4), 253–260.

Hartmanis, J., and Hopcroft, J.E. 1976. Independence results in computer science. *SIGACT News,* **8**(4), 13–24.

Heller, Alex. 1990. An existence theorem for recursion categories. *J. Symbolic Logic,* **55**(3), 1252–1268.

Hodges, Andrew. 1983. *Alan Turing: the Enigma.* Simon & Schuster.

Hofstra, Pieter, and Scott, Philip. 2021. Aspects of categorical recursion theory. Downloaded from https://www.site.uottawa.ca/ phil/papers on December 24, 2019.

Hopcroft, John E., and Ullman, Jeff D. 1979. *Introduction to Automata Theory, Languages, and Computation.* Addison-Wesley.

Immerman, Neil. 1998. *Descriptive Complexity.* Springer.

Kassel, Christian. 1995. *Quantum Groups.* Graduate Texts in Mathematics, vol. 155. Springer.

Lambek, Joachim, and Scott, Philip. 1986. *Introduction to Higher-Order Categorical Logic.* Cambridge Studies in Advanced Mathematics, no. 7. Cambridge University Press.

Lawvere, F. William. 1969. Diagonal arguments and cartesian closed categories. Pages 134–145 of: *Category Theory, Homology Theory and their Applications, II (Battelle Institute Conference, Seattle, Wash., 1968, Vol. Two).* Springer.

Lawvere, F. William, and Rosebrugh, Robert. 2003. *Sets for Mathematics.* Cambridge University Press.

Lawvere, F. William, and Schanuel, Stephen H. 2009. *Conceptual Mathematics: A First Introduction to Categories,* Second edition. Cambridge University Press.

Leinster, Tom. 1998. Basic bicategories. Available at arXiv:math/9810017.

Lengyel, Florian. 2004. More existence theorems for recursion categories. *Ann. Pure Appl. Logic,* **125**(1–3), 1–41.

Lewis, Harry R., and Papadimitriou, Christos H. 1997. *Elements of the Theory of Computation,* Second edition. Prentice Hall.

Libkin, Leonid. 2004. *Elements of Finite Model Theory.* Springer.

Longo, Giuseppe, and Moggi, Eugenio. 1984a. Cartesian closed categories of enumerations for effective type structures (Part I & II). Pages 235–255 in: *Semantics of Data Types*, Khan, G., MacQueen, D.B., and Plotkin, G. (eds) Lecture Notes in Computer Science, vol. 173. Springer.

Longo, Giuseppe, and Moggi, Eugenio. 1984b. Gödel numberings, principal morphisms, combinatory algebras: A category-theoretic characterization of functional completeness. Pages 397–406 in: *Mathematical Foundations of Computer Science 1984. MFCS 1984*, Chytil M.P., Koubek V. (eds). Lecture Notes in Computer Science, vol. 176. Springer.

Longo, Giuseppe, and Moggi, Eugenio. 1990. A category-theoretic characterization of functional completeness. *Theor. Comput. Sci.*, **70**(2), 193–211.

Mac Lane, Saunders. 1998. *Categories for the Working Mathematician*, Second edition. Graduate Texts in Mathematics, vol. 5. Springer.

Manin, Yu. I. 2010. *A Course in Mathematical Logic for Mathematicians*, Second edition. Graduate Texts in Mathematics, vol. 53. Springer.

Manin, Yuri I. 2000. Classical computing, quantum computing, and Shor's factoring algorithm. *Astérisque*, **266**, 375–404.

Mendelson, Elliott. 1997. *Introduction to Mathematical Logic*, Fourth edition. Chapman & Hall.

Otto, James R. 1995. *Complexity Doctrines*. Ph.D. Thesis, McGill University.

Papadimitriou, Christos H. 1994. *Computational Complexity*. Addison-Wesley.

Pavlovic, Dusko. 2013. Monoidal computer I: Basic computability by string diagrams. *Information and Computation*, **226**(Supplement C), 94–116.

Pierce, Benjamin C. 1991. *Basic Category Theory for Computer Scientists*. MIT Press.

Poigné, Axel. 1992. Basic Category Theory. Pages 413–640 of: *Handbook of Logic in Computer Science (Vol. 1)*, Abramsky, S., and Maibaum, T.S.E. (eds). Oxford University Press, Inc.

Rich, Elaine. 2008. *Automata, Computability and Complexity: Theory and Applications*. Pearson.

Rogers, Jr., Hartley. 1987. *Theory of Recursive Functions and Effective Computability*. MIT Press.

Román, Leopoldo. 1989. Cartesian categories with natural numbers object. *J. Pure Appl. Algebra*, **58**(3), 267–278.

Rydeheard, David E., and Burstall, Rod M. 1988. *Computational Category Theory*. Prentice Hall.

Scott, P.J. 2000. Some aspects of categories in computer science. Pages 3–77 of: *Handbook of Algebra, Vol. 2*, Hazewinkel, M. (ed). North-Holland.

Seely, R.A.G. 1987. Modeling computations: a 2-categorical framework. Pages 65–71 of: *Proc. Symposium on Logic in Computer Science*, IEEE Computer Society,

Shannon, Claude E. 1956. A universal Turing machine with two internal states. Pages 157–165 of: *Automata Studies*. Annals of Mathematics Studies, no. 34. Princeton University Press.

Shor, Peter. W. 1994. Algorithms for quantum computation: discrete logarithms and factoring. Pages 124–134 of: *Proc. 35th Annual Symposium on Foundations of Computer Science.* IEEE Computer Society.

Shulman, Michael A. 2010. Constructing symmetric monoidal bicategories. Available at arXiv:1004.0993.

Sipser, Michael. 2006. *Introduction to the Theory of Computation,* Second edition. Course Technology.

Soare, Robert I. 1987. *Recursively Enumerable Sets and Degrees: A Study of Computable Functions and Computably Generated Sets.* Springer.

Spivak, David I. 2014. *Category Theory for the Sciences.* MIT Press.

Street, Ross. 1996. Categorical structures. Pages 529–577 of: *Handbook of Algebra, Vol. 1.* Elsevier/North-Holland.

Turing, Alan. 1937. On computable numbers with an application to the Entscheidungsproblem. *Proceedings of the London Mathematical Society (2),* **42**, 230–265.

Univalent Foundations Program. 2013. *Homotopy Type Theory – Univalent Foundations of Mathematics.* https://homotopytypetheory.org/book, Institute for Advanced Study.

Walters, R.F.C. 1991. *Categories and Computer Science.* Cambridge University Press.

Yanofsky, Noson S. 2003. A universal approach to self-referential paradoxes, incompleteness and fixed points. *Bull. Symbolic Logic,* **9**(3), 362–386.

Yanofsky, Noson S. 2011. Towards a definition of an algorithm. *J. Logic Comput.,* **21**(2), 253–286.

Yanofsky, Noson S. 2013. *The Outer Limits of Reason: What Science, Mathematics, and Logic Cannot Tell Us.* MIT Press.

Yanofsky, Noson S. 2015. Computability and Complexity of Categorical Structures. Available at http://www.sci.brooklyn.cuny.edu/noson/.

Yanofsky, Noson S. 2017. Galois theory of algorithms. Pages 323–347 of: *Rohit Parikh on Logic, Language and Society,* Başkent, C., Moss, L., Ramanujam R. (eds). Outstanding Contributions to Logic, vol. 11. Springer.

Yanofsky, Noson S., and Mannucci, Mirco A. 2008. *Quantum Computing for Computer Scientists.* Cambridge University Press.

Acknowledgements

There are many people who made this Element possible. In 1989, as an undergraduate at Brooklyn College, I had the privilege of taking a master's level course in theoretical computer science with Rohit Parikh. This world-class expert made the entire subject come alive, and I have been hooked ever since. Although thirty years has passed, I am still amazed at how much he knows and – more importantly – how much he understands on a deeper level. He was my teacher, and now he is a colleague and a friend. I am forever grateful to him.

One of the founders of category theory was Saunders Mac Lane (1909–2005). Whenever we met at a conference, he was always warm and friendly. He wrote one of the "bibles" of the field, *Categories for the Working Mathematician* (Mac Lane, 1998). As the title implies, the Element teaches category theory to someone who is well-grounded in mathematics. In 1999, I wrote him a letter about a book I wanted to write that was supposed to teach much advanced mathematics to someone who is well-grounded in basic category theory. I asked him for permission to call the book *Mathematics for the Working Category Theorist*. Saunders was encouraging and gave his imprimatur. What you are reading now is a first step towards that dream.

I would like to thank John Baez for taking an interest in this work and for his support. Over the years, I gained so much from John's amazing papers and posts. His clarity and lucidity are an inspiration. More than anyone else, John has shown how the language of category theory can be applied in so many different areas. In addition, Yuri Manin's paper (Manin, 2000) and book (Manin, 2010) have been an inspiration for this Element. He put all computational processes into one category called a "computational universe." Here we take these ideas further. I thank him for his ideas and for taking an interest in my work.

Ted Brown, the former chair of the computer science department at The Graduate Center of the City University of New York, has been very kind to me. He asked me to teach the PhD level theoretical computer science course every year from 2003 through 2013. I also taught several other advanced courses in those years. That experience has been tremendously helpful. My thanks also goes out to the next chair of the department, Robert M. Haralick. Their kindness is appreciated.

Brooklyn College has been my intellectual home since I was an entering freshman in 1985. The entire administration, faculty, staff, and students have been encouraging and have made the environment conducive to such projects. Dean Kleanthis Psarris, Chairman Yedidyah Langsam, and former Chairman Aaron Tenenbaum were all very helpful. Over the years I also gained much from David Arnow, Eva Cogan, Lawrence Goetz, Ira Rudowsky, and Joseph Thurm. I thank them all.

The following were helpful with discussions, encouragement, and editing: Sergei Artemov, Gershom Bazerman, Chris Calude, John Connor, James Cox, Walter Dean, Scott Dexter, Mel Fitting, Grant Roy, Tzipora Halevi, Joel David Hamkins, Keith Harrow, Pieter Hofstra, Karen Kletter, Roman Kossak, Deric Kwok, Moshe Lach, Florian Lengyel, Armando Matos, Michael Mandel, Yuri Manin, Jean-Pierre Marquis, Robert Paré, Rohit Parikh, Vaughn Pratt, Avi Rabinowitz, Phil Scott, Robert Seely, Morgan Sinclaire, David Spivak, Alexander Sverdlov, Gerald Weiss, Paula Whitlock, Mark Zelcer, and all the members of The New York City Category Theory Seminar.

Bob Coecke, Joshua Tan and David Tranah have been great editors and have shepherded this work along from the beginning. I am also grateful to two anonymous reviewers for their many comments and helpful suggestions. Thank you!

This work was done with the benefit of the generous support of PSC-CUNY Award 61522-00 49.

Neither this Element nor anything else could have been done without my wife Shayna Leah. Her loving help in every aspect of my life is deeply appreciated. I am grateful to my children, Hadassah, Rivka, Boruch, and Miriam for all the joy they bring me.

Index

Ackermann, Wilhelm, 42
algebraic theory, 100
algorithm, 3, 4, 8, 85, 92, 94, 98, 126
Artemov, Sergei, 135
associator, 15
automata, cellular, 50
automata, finite, 10, 126

Bénabou, Jean, 14, 16
Baez, John, 134
bicategory, 14, 29
Brown, Ted, 134

c.e., 61
Calude, Christian, 135
Cantor, Georg, 2, 7, 104, 107
Cartesian closed categories, 60, 106
categorical imperative, 128
Church, Alonzo, 33, 50, 73
circuits, 6, 43–51
coalgebra, 50
Cockett, Robin, 129
Coecke, Bob, 135
coherence, 12–16, 20, 29
comma category, 11, 16, 67, 91, 102, 126
complexity class, 88, 89, 91, 99, 100, 102, 126
complexity measure, 77–90
complexity theory, 2, 3, 7, 26, 77–104
computability class, 67
computability theory, 2, 3, 7, 60–77
computably enumerable, 61
computationally independent, 76
coslice category, 9–11
creative sets, 74
cryptography, 3, 8
currying, 60

Davis, Martin, 74
Dean, Walter, 103, 135
descriptive complexity theory, 56, 101
DEXPSPACE, 103
diagonal argument, 2, 7, 8, 104–125
Diaz-Boils, Ximo, 129
Dijkstra, Edsger, 32
Diophantine equations, 74
DiPaola, Robert, 129
dominical categories, 129

Eilenberg, Samuel, 4, 129
Elgot, Calvin C., 129
encoding, 56–60
Entscheidungsproblem, 73, 127
Epimenides, 7

Euler cycle, 92
Euler, Leonhard, 92
existential quantifier, 101

finite model theory, 56, 101
Fitting, Mel, 135
fixed point, 108
formal language theory, 3, 8
Friedberg, Richard, 76
function, Ackermann, 43, 114
function, basic, 38
function, computable, 16–21, 61
function, decidable, 61
function, exponential, 80, 85, 89, 93, 103
function, factorial, 39, 80, 85
function, pairing, 56
function, partially computable, 61
function, polynomial, 80
function, primitive recursive, 33–43, 114
function, projection, 38
function, recursive, 33–43, 61
function, solvable, 61
function, space-constructable, 118
function, successor, 38
function, time-constructable, 118
function, timed halting, 64
function, total computable, 16–21, 61
function, totally solvable, 61
function, totally Turing computable, 61
function, transition, 25, 28, 86, 87
function, Turing computable, 61
function, Turing decidable, 61
function, unary, 114
function, zero, 38
functor, adjoint, 126
functor, essentially surjective, 24
functor, inherently surjective, 24, 33, 44
functor, surjective, 127

Gödel numbering, 58
Gödel, Kurt, 1, 2, 7, 33
Gödelization, 58
graph, finite, 10

Hamkins, Joel David, 135
Haralick, Robert M., 134
Harel, David, 128
Harrow, Keith, 128
Hartmanis, Juris, 124
Heller, Alex, 129
Hilbert, David, 42, 73, 74
Hofstra, Pieter, 129, 135
homotopy type theory, 20

Hopcroft, John E., 124
hyperimmune sets, 74
hypersimple sets, 74

identity constraints, 15
Immerman, Neil, 102
immune sets, 74

jump operation, 76

Kan extension, 82, 83, 122, 126
Kleene, Stephen, 2, 33
Kolmogorov complexity theory, 3, 8
Kossak, Roman, 135

Ladner, Richard E., 100
lambda calculus, 20
Lambek, Joachim, 33
Leinster, Tom, 16
Lengyel, Florian, 129, 135
liar paradox, 62
log-space transducers, 46, 102
logic of decision problems, 101
logical formula, 51–56
logical system, complete, 71
logical system, consistent, 71
logical system, incomplete, 71
logical system, inconsistent, 71
Longo, Giuseppe, 129

Mac Lane, Saunders, 4, 12, 134
machine, Post, 50
machine, self-replicating, 2, 118
Manin, Yuri, 40, 129, 134, 135
Marquis, Jean-Pierre, 135
Matiyasevich, Yuri, 74
Minsky, Marvin, 33
models of computation, 2, 6, 16–60
Moggi, Eugenio, 129
monad, 100
monoidal category, 5, 12
Muchnik, Albert, 76

natural number object, 41, 42
natural number object, parameterized, 42
NEXPSPACE, 103
NP, 91, 95
NPSPACE, 102

operad, 100
operation, μ-minimization, 38, 41
operation, composition, 38
operation, jump, 76
operation, recursion, 38
Otto, Jim, 129

P, 91, 95
P-categories, 129

P=NP question, 1, 2, 98, 100, 121, 124
Paré, Robert, 135
Parikh, Rohit, 134, 135
Pavlovic, Dusko, 129
platonism, 51, 53
Poincaré, Henri, 118
Post, Emile, 50
post-quantum cryptography, 4
Pratt, Vaughn, 135
problem coPath, 99
problem, coEuler, 99
problem, coNP, 99
problem, coP, 99
problem, coPrime, 99
problem, decision, 9, 23, 60–77, 90–101
problem, Discrete logarithm, 92, 93
problem, Empty program, 68
problem, Equivalent program, 68
problem, Euler cycle, 92
problem, Factoring, 92, 93
problem, Graph isomorphism, 92
problem, Halting, 2, 60–77, 111, 118
problem, Hamiltonian cycle, 94–96
problem, Hilbert's 10th, 74
problem, intractable, 1
problem, Königsberg bridge, 92
problem, Knapsack, 93, 96
problem, L, 102
problem, L-complete, 102
problem, NL, 102
problem, NL-complete, 102
problem, Nonempty program, 67
problem, NP, 91–100
problem, NP-complete, 2, 97–99, 104
problem, NP-intermediate, 98, 100
problem, NPSPACE-complete, 102
problem, Path, 92
problem, Prime, 92
problem, Printing 42, 69
problem, PSPACE-complete, 102
problem, Satisfiability, 85, 87, 93, 99, 126
problem, Set partition, 94
problem, Subset sum, 94–96, 99
problem, Traveling salesman, 85, 94–96
problem, unHamilton, 99
problem, unSatisfiable, 99
problem, unsolvable, 1
problem, unSubset sum, 99
problem. Halting, 126
productive sets, 74
protocol, cryptographic, 8, 126
PSPACE, 102
Putnam, Hilary, 74

quantum computers, 4, 98, 129

r.e., 61, 74
Rabinowitz, Avi, 135

realism, 51, 53
recursion categories, 129
recursive sets, 74
recursively enumerable, 61
reduction, 10, 11, 66, 95
reduction, polynomial, 95
register machine, 6, 33–43
representable function, 108
restriction categories, 129
Robinson, Julia, 74
robust, 32, 34
Román, Leopoldo, 42
Russell, Bertrand, 7

Savitch, Walter, 102
Scott, Philip, 129, 135
search space, 100
Seely, Robert, 129, 135
semantics, 7, 8, 70
Shannon, Claude, E., 32
Shepherdson, John C., 33
simple sets, 74
slice category, 9–11
space complexity, 101–104
Spivak, David, 23, 135
Street, Ross, 16
strict monoidal category, 12
strictly associative monoidal category, 12
strong symmetric monoidal functor, 13
symmetric monoidal bicategory, 12–16
symmetric monoidal category, 6, 12–16
syntax, 7, 8, 70
Szelepcsényi, Róbert, 102

Tan, Joshua, 135
terminal object, 97
terminal object, weak, 96, 97, 126
The Big Picture, 16–21
theorem, Baker–Gill–Solovay, 121, 124, 125
theorem, Cantor's, 104–125
theorem, contrapositive of Cantor's, 110–125
theorem, Cook–Levin, 97, 127
theorem, Friedberg–Muchnik, 76, 100
theorem, fundamental, of calculus, 21
theorem, Gödel's incompleteness, 1, 2, 71, 72,
    104, 118, 127
theorem, generalized Ladner, 100
theorem, hairy ball, 118
theorem, Immerman–Szelepcsényi, 102
theorem, Kleene normal form, 41
theorem, Ladner, 100, 124
theorem, nondeterministic space hierarchy, 120
theorem, nondeterministic time hierarchy, 121
theorem, recursion, 2, 116, 126
theorem, Rice's, 70, 71, 117
theorem, Rogers' version of Kleene's recursion,
    116
theorem, Savitch's, 102, 122, 127

theorem, second recursion, 116
theorem, SMN, 59, 117
theorem, space hierarchy, 119
theorem, time hierarchy, 120
theorem, undecidability of the Halting problem,
    62
thesis, Church–Turing, 31, 32, 37, 40, 50, 62
Turing categories, 129
Turing machine, 6, 21–33
Turing machine, alternating, 89
Turing machine, deterministic, 32, 85–89, 127
Turing machine, nondeterministic, 32, 85–90,
    127
Turing machine, oracle, 75–77, 121–124
Turing machine, probabilistic, 89
Turing machine, quantum, 89
Turing machine, reversible, 89
Turing machine, universal, 59
Turing, Alan, 1, 2, 4, 7, 25, 60, 73, 75, 104
type theory, 19
type, function, 18
type, list, 18
type, product, 18
type, sequence of, 19
types, 16–21
types, basic, 18

universal quantifier, 101

von Neumann, John, 2, 118

Wang, Hao, 33
weak Lindenbaum–Tarski category, 55

Zermelo–Frankel axioms, 124

Cambridge Elements ☰

Elements in Applied Category Theory

Bob Coecke

Cambridge Quantum Ltd

Bob Coecke is Chief Scientist at Cambridge Quantum Ltd. and Emeritus Professor at Wolfson College, University of Oxford. His pioneering research includes categorical quantum mechanics, ZX-calculus, quantum causality, resource theories, and quantum natural language processing. Other interests include cognitive architectures and diagrams in education. Most of his work uses the language of tensor categories and their diagrammatic representations. He co-authored the book Picturing Quantum Processes. A First Course in Quantum Theory and Diagrammatic Reasoning. He is considered as one of the fathers of the field of applied category theory, and a co-founder of the journal Compositionality.

Joshua Tan

University of Oxford

Joshua Tan is a doctoral student at the University of Oxford and a Practitioner Fellow at Stanford University. He is the executive director of the Metagovernance Project and an executive editor of the journal Compositionality. He works on applications of category theory and sheaf theory to theoretical machine learning and to the design of complex, multi-agent systems.

About the Series

Elements in Applied Category Theory features Elements intended both for mathematicians familiar with category theory and seeking elegant, graduate-level introductions to other fields in the language of categories, and for subject-matter experts outside of pure mathematics interested in applications of category theory to their field.

Cambridge Elements $\equiv$

Elements in Applied Category Theory

Elements in the Series

Theoretical Computer Science for the Working Category Theorist
Noson S. Yanofsky

A full series listing is available at: www.cambridge.org/EACT

Printed in the United States
by Baker & Taylor Publisher Services